Transactions on Computational Science and Computational Intelligence

Series Editor
Hamid Arabnia

Associate Editor
Pietro Cipresso

More information about this series at http://www.springer.com/series/11769

Babak Akhgar • Andrew Staniforth
David Waddington

Editors

Application of Social Media
in Crisis Management

Advanced Sciences and Technologies
for Security Applications

 Springer

Editors
Babak Akhgar
CENTRIC
Cultural, Communication and Computing
    Research Institute (C3RI)
Sheffield Hallam University
Sheffield, UK

Andrew Staniforth
West Yorkshire Police
Wakefield, UK

David Waddington
CENTRIC
Cultural, Communication and Computing
    Research Institute (C3RI)
Sheffield Hallam University
Sheffield, UK

Transactions on Computational Science and Computational Intelligence
ISBN 978-3-319-84901-0   ISBN 978-3-319-52419-1   (eBook)
DOI 10.1007/978-3-319-52419-1

Printed on acid-free paper

This Springer imprint is published by Springer Nature
The registered company is Springer International Publishing AG
The registered company address is: Gewerbestrasse 11, 6330 Cham, Switzerland

Acknowledgments

The editors would like to take this opportunity to thank the multidisciplinary team of contributors who dedicated their time, knowledge and experiences in preparing the chapters contained in this edited volume. In particular, we would like to recognise the dedication of Stephanie Young (of the Swedish Defense University) and the wider team at CENTRIC (Centre of Excellence in Terrorism, Resilience, Intelligence and Organised Crime Research, Sheffield Hallam University) without whom this edited volume would not have been possible.

We also extend our thanks to the consortium partners and advisory board members of the EU FP7 ATHENA project for the encouragement and support they have given, not only to the writing of this book but to the project as a whole, and for the co-ordination and commitment displayed by ATHENA Project Manager, Jessica Gibson (of the Office of the Police and Crime Commissioner for West Yorkshire).

The full list of consortium partners to whom we are so greatly indebted is as follows:

- West Yorkshire Police (United Kingdom)
- International Organisation for Migration (Belgium)
- Sheffield Hallam University (United Kingdom)
- Fraunhofer Institute (Germany)
- SAS Software Limited (United Kingdom)
- Municipality of Ljubljana (Slovenia)
- Thales Nederland BV (Netherlands)
- University of Virginia (United States)
- Försvarshögskolan - Swedish Defense University (Sweden)
- EPAM Systems (Sweden)
- Izmir Buyuksehir Belediyesi (Turkey)
- Research in Motion Limited (Canada)
- Epidemico Limited (Ireland)
- Police National Legal Database (United Kingdom)

Finally, we would like to gratefully acknowledge that the ATHENA project was made possible due to funding received from the European Union's Seventh Framework Programme for research, technological development and demonstration (FP7-SEC-2013) under grant agreement number 313220.

Contents

Editors' Biographies

Babak Akhgar, Ph.D., F.B.C.S. is Professor of Informatics and Director of CENTRIC (Centre of Excellence in Terrorism, Resilience, Intelligence and Organised Crime Research) at Sheffield Hallam University (SHU). A Fellow of the British Computer Society, he has more than 100 referred publications in international journals and conferences on information systems, with a specific focus on knowledge management (KM). He is a member of the editorial boards of a number of international journals and chair and programme committee member of several international conferences. He has extensive and hands-on experience in the development, management and execution of large international security initiatives (e.g. application of social media in crisis management, intelligence-based combating of terrorism and organised crime, gun crime, cyber security, Big Data and cross-cultural ideology polarisation) with multimillion Euro budgets. In addition to this, he is currently the technical lead of three EU Security projects: 'Courage' on Cyber Crime and Cyber Terrorism, 'Athena project' on the application of social media and mobile devices in crisis management and TENSOR on the identification of terrorist-generated Internet content. He has co-edited a book on intelligence management titled *Knowledge Driven Frameworks for Combating Terrorism and Organised Crime* (with Simeon Yates, Springer, 2011). His recent books are titled *Strategic Intelligence Management—National Security Imperatives and Information and communication Technologies* (with Simeon Yates, Elsevier, 2013), *Information and Communications Technologies Emerging Trends in ICT Security* (with Hamid Arabnia, Elsevier, 2013), *Application of Big Data for National Security—A Practitioners Guide to Emerging Technologies* (with Gregory B. Saathoff, Hamid Arabnia, Richard Hill, Andrew Staniforth and Petra Bayerl, Elsevier, 2015) and also *Open Source Intelligence Investigation—From Strategy to Implementation* (with Petra Bayerl and Fraser Sampson, Springer, 2016). He is also a board member of European Organisation for Security (EOS) and member of the academic board of SAS UK.

Andrew Staniforth, Ph.D. is a Detective Inspector in the West Yorkshire Police force, who has extensive counterterrorism experience in the UK. As a professionally qualified teacher, he has designed national counterterrorism training and exercising programmes, delivered training to police commanders from across the world and supported missions of the United Nations Terrorism Prevention Branch. He is a Senior Research Fellow at the School of Law, University of Leeds, and a Non-Resident Fellow in Counter-Terrorism and National Security at the Trends Research and Advisory Institute. He is the author and editor of numerous articles on counter-terrorism and national security. His most recent books include: *Blackstone's Counter-Terrorism Handbook (3rd edition)* (Oxford University Press 2013); *Blackstone's Handbook of Ports and Borders Security* (Oxford University Press 2013); *Preventing Terrorism and Violent Extremism* (Oxford University Press 2014); *Cyber Crime and Cyber Terrorism Investigators Handbook* (Elsevier, 2014); *Blackstone's Handbook of Cyber Crime Investigation* (Oxford University Press in press 2017); and *Big Data Applications for National Security* (Elsevier 2015). He now leads an innovative police research team at the Office of the Police and Crime Commissioner for West Yorkshire, progressing multidisciplinary international research and innovation projects. Andrew the Project Coordinator of ATHENA, funded by the European Commission under the Seventh Framework Programme.

David Waddington, Ph.D. is Professor of Communications, Co-director of the Cultural, Communication and Computing Research Institute, Academic Chair of CENTRIC (Centre of Excellence in Terrorism, Resilience, Intelligence and Organised Crime Research) and Head of the Communication and Computing Research Centre at Sheffield Hallam University (SHU). He was also Chair of his faculty's Research Ethics Committee from February 2008 to February 2013. He has been employed at SHU (which was previously known as Sheffield City Polytechnic) since 1983—initially as a Postdoctoral Research Associate on an ESRC project investigating 'Communication processes within and around "flashpoints" of Public Disorder'. This focus on the policing of riots, disorderly demonstrations and picket-line confrontations was instrumental to the development of his 'Flashpoints Model of Public Disorder', which is frequently referred to in the European, North American and Antipodean policing literatures. Among his best-known publications are the seminal *Flashpoints: Studies in Public Disorder* (with Chas Critcher and Karen Jones, Routledge, 1989); *Contemporary Issues in Public Disorder* (Routledge, 1992); *Public Order Policing: Theoretical and Practical Approaches* (Willan, 2007); *Rioting in France and the UK: A Comparative Analysis* (co-edited with Fabien Jobard and Mike King, Willan, 2009); and *Riots—An International Comparison* (with Matthew Moran, Palgrave Macmillan, 2016). He is currently on the International Editorial Board of *Mobilization: An International Social Movements Journal*. From 2011 to 2013. He was External Evaluator of the EU-funded GODIAC project, which brought together 12 European partner countries in search of a distinctively more enlightened and permissive 'European approach' to protest policing.

Authors' Biographies

Abdelhaq Abouhafc has a Masters degree in Applied Mathematics from Delft University of Technology. Since August 2014, he has worked at Thales as a research software engineer. He is working on the Dynamic Process Integration Framework (DPIF) project. His research interest focuses on algorithms development.

Babak Akhgar is Professor of Informatics and Director of CENTRIC (Centre of Excellence in Terrorism, Resilience, Intelligence and Organised Crime Research) at Sheffield Hallam University, and Fellow of the British Computer Society.

Simon Andrews is Professor of Conceptual Structures in Computer Science at Sheffield Hallam University, where he develops algorithms and applications for Formal Concept Analysis, a mathematical approach to the analysis of object/attribute data. He has been Principal Investigator on several major EU projects and is a lead researcher on the ATHENA project.

Chi Bahk is Vice President of Operations at Epidemico, a public health technology company specialising in utilising nontraditional data sources, such as online news, social media and crowdsourcing for health intelligence. Chi's background is in public health, and at Epidemico, she leads both government and commercial projects, including multinational research projects.

Lucas Baptista is a Mobile Software Engineer at Epidemico, with 5 years' experience on mobile development and background on backend development.

Kevin Blair is a Project Manager with extensive experience across law enforcement and government, delivering national and EU-funded projects. He specialises in technology development and deployment. Recent projects have taken him back to his roots in developing policy and processes, at a national level, for UK law enforcement.

Robin Colodzin is a consultant with over 20 years of experience as a backend web developer.

Ravi Coote studied computational linguistics and computer science at the University of Bonn and graduated in 2009 with a thesis in speech recognition. He is employed as a research assistant at Fraunhofer FKIE, supporting various national and international research projects such as crisis management, smart cities and cybercrime.

Tony Day is a researcher within CETRIC at Sheffield Hallam University. His primary research interest is the application of cutting-edge and open-source technologies in areas with security, humanitarian and environmental impacts.

Konstantinos Domdouzis is a researcher within CENTRIC with a background in Computer Science and Computer Networks and Communications. His Ph.D. work focused on the applications of Wireless Sensor Technologies in the Construction Industry and his postdoctoral research focused on Systems Informatics for Biomass Feedstock Production.

Andrej Fink has served since 1994 as Head of University Medical Centre Ljubljana Ambulance Service. He is intensively involved in the development of Emergency Medical Service in Slovenia. He is also a lecturer at Faculty of Health Care Jesenice and College of Health Studies, University of Maribor. He holds a master's degree in Health Sciences—Emergency Services Management.

Laurence Hirsch works in the Computing Department at Sheffield Hallam University and teaches in various computing subjects including web programming, cloud systems, database development and machine learning. His research specialises in text visualisation and evolutionary computation, applied to text classification and clustering.

Julij Jeraj has served as a senior advisor in disaster management for the City of Ljubljana since 1996. In this position, he conducts vulnerability and risk assessment, emergency planning, research and development, international cooperation, disaster and emergency management. He holds a Masters degree in security studies and political science.

Raluca Lefticaru holds a Ph.D. and Masters degree in Computer Science and is a researcher within CENTRIC at Sheffield Hallam University.

Alison Lyle has been with West Yorkshire Police for 11 years and is currently a member of the West Yorkshire for Innovation Team working on EU-funded projects. Specialises in legal work relating to data protection, police law (UK) and cybercrime. She represents the Police National Legal Database on Project ATHENA, fulfilling the role of Legal Adviser.

Kerry McSeveny is a researcher and associate lecturer in the Communication and Computing Research Centre (CCRC) at Sheffield Hallam University. She holds a B.A. in Psychology, an M.A. in Communication Studies and a Ph.D. on the management of identity online. She is interested in online identity and community and how social 'issues' manifest and are represented in cultural discourse in interaction, social media and mass media.

Constantinos Orphanides is a researcher in CENTRIC, specialising in software engineering and scaling data for Formal Concept Analysis (FCA) and the author of software to automate this task. He is a member of the Conceptual Structures Research Group where he is completing his Ph.D. titled 'Appropriating Data from Structured Sources for Formal Concept Analysis'.

Patrick de Oude is senior researcher at Thales Nederland B.V. He received his Masters and Ph.D. degrees in Artificial Intelligence at the University of Amsterdam. His interests and work are centred around (distributed) probabilistic reasoning and modelling, graphical models, causality, multi-agent systems, data mining and machine learning.

Gregor Pravlin is the manager at Thales Nederland B.V. He received his Ph.D. in computer science from Graz University of Technology, Austria. He has extensive industrial experience in the development of complex, safety critical airborne software systems. He is also a visiting researcher at the Intelligent Autonomous Systems group at the University of Amsterdam.

Thomas Quillinan is a Security Researcher at the D-CIS Research Lab in Delft, in the Netherlands. He received a Ph.D., in the area of Security for Distributed Systems, and an M.Sc. in Computer Science from University College Cork in Ireland. His postdoctoral research focused on Distributed Systems and Crisis Management.

Kellyn Rein is a Research Associate at the Fraunhofer Gesellschaft, Europe's largest applied research organisation. Her area of specialty is in the analysis of lexical forms of uncertainty in natural language information. She is also involved in numerous NATO Research Task Groups, chairing one on multilevel, multi-source information fusion.

Fraser Sampson is Chief Executive and Solicitor of the Office of Police and Crime Commissioner West Yorkshire Police. Before taking up the role, he was Chief Executive and Solicitor of the West Yorkshire Police Authority, and in 2008 he was the first Executive Director of the Civil Nuclear Police Authority created by the Energy Act 2004.

Ulrich Schade is a senior research scientist and head of the research group 'Information Analysis' with Fraunhofer FKIE. He is also Associate Professor to Bonn University, teaching 'Applied Linguistics'.

Lukas Sikorski is a Research Associate at Fraunhofer Institute for Communication, Information Processing and Ergonomics (FKIE). He holds a diploma in Applied Computer Science. At present, he is working on building a domain-specific ontology containing knowledge about the energy domain, as well as supporting multilingual support for CML.

Andrew Staniforth is a Senior Research Fellow and Advisory Board Member of CENTRIC. After becoming a qualified reader and published author, he is now a Police Detective Inspector, combining his practical expertise with applied research in the security domain.

Eric K. Stern is Professor at the College of Emergency Preparedness, Homeland Security, and Cyber-Security at the University at Albany (SUNY). He is also affiliated with the Center for Crisis Management Research and Training (CRISMART at the Swedish Defense University), where he served as Director from 2004 to 2011.

David Waddington is Professor of Communications, Co-director of the Cultural, Computing and Communications Research Institute and Head of the Communication and Computing Research Centre at Sheffield Hallam University.

Carly Winokur is a Senior Marketing Consultant, experienced in establishing effective health communications initiatives. She is interested in making public health palatable and the utilisation of nontraditional data to complement traditional methods. She has an M.P.H. from Emory University and a B.S. in International Relations/Communication Studies from the University of Miami.

Contributors

Abdelhaq Abouhafc Thales Research and Technology Netherlands, Delft, The Netherlands

Babak Akhgar CENTRIC, Cultural, Communication and Computing Research Institute (C3RI), Sheffield Hallam University, Sheffield, UK

Simon Andrews CENTRIC, Sheffield Hallam University, Sheffield, UK

Chi Bahk Epidemico, Boston, Massachusetts, USA

Lucas Baptista Epidemico, Boston, Massachusetts, USA

Kevin Blair Office of the Police and Crime Commissioner for West Yorkshire, Wakefield, UK

Robin Colodzin Epidemico, Boston, Massachusetts, USA

Ravi Coote Fraunhofer FKIE, Wachtberg, Germany

Tony Day CENTRIC, Sheffield Hallam University, Sheffield, UK

Konstantinos Domdouzis CENTRIC, Sheffield Hallam University, Sheffield, UK

Andrej Fink The City of Ljubljana, Emergency Management Department, Ljubljana, Slovenia

Laurence Hirsch CENTRIC, Sheffield Hallam University, Sheffield, UK

Julij Jeraj The City of Ljubljana, Emergency Management Department, Ljubljana, Slovenia

Raluca Lefticaru CENTRIC, Sheffield Hallam University, Sheffield, UK

Alison Lyle CENTRIC, Sheffield Hallam University, Sheffield, UK

Kerry McSeveny CENTRIC, Sheffield Hallam University, Sheffield, UK

Constantinos Orphanides CENTRIC, Sheffield Hallam University, Sheffield, UK

Patrick de Oude Thales Research and Technology Netherlands, Delft, The Netherlands

Gregor Pavlin Thales Research and Technology Netherlands, Delft, The Netherlands

Thomas Quillinan Thales Research and Technology Netherlands, Delft, The Netherlands

Kellyn Rein Fraunhofer FKIE, Wachtberg, Germany

Fraser Sampson Office of the Police and Crime Commissioner for West Yorkshire, UK

Ulrich Schade Fraunhofer FKIE, Wachtberg, Germany

Lukas Sikorski Fraunhofer FKIE, Wachtberg, Germany

Andrew Staniforth West Yorkshire Police, Wakefield, UK

Eric K. Stern CRiSMART, Swedish Defense University, Stockholm, Sweden

University at Albany, SUNY, Albany, NY, USA

David Waddington CENTRIC, Cultural, Communication and Computing Research Institute (C3RI), Sheffield Hallam University, Sheffield, UK

Carly Winokur Epidemico, Boston, Massachusetts, USA

Chapter 1
Introduction

Babak Akhgar, Andrew Staniforth, and David Waddington

1.1 Scope and Primary Objective

The primary objective of this volume is to show, by focusing on one particular example of specially customised digital system, how the vital (and potentially life-saving) sense-making activities involved in the management of crises and disasters can be considerably enhanced via processes of crowdsourcing and communication facilitated by social media and personal communications technology.

The contributing chapters to this book will be used collectively to demonstrate how socio-technical platforms of this nature have the capacity to enable multidirectional exchange of information and provide windows into the perceptions, predispositions, and concerns (short term and longer term) of citizens, police, and other emergency services. It will become evident from the contents of our chapters that social media-based information can serve as a complement to other, more 'traditional' sources of information/intelligence; and that, finding themselves empowered by their ability to gain access to personal communications devices and networks, individuals are now able to document and share (potentially time-stamped and geo-tagged) text, images, and video in ways that complement more conventional forms of reporting.

B. Akhgar (✉) • D. Waddington
CENTRIC, Cultural, Communication and Computing Research Institute (C3RI), Sheffield Hallam University, Sheffield, UK
e-mail: B.Akhgar@shu.ac.uk; D.P.Waddington@shu.ac.uk

A. Staniforth
West Yorkshire Police, Wakefiled, UK
e-mail: andrew.staniforth1@westyorkshire.pnn.police.uk

© Springer International Publishing AG 2017   1
B. Akhgar et al. (eds.), *Application of Social Media in Crisis Management*,
Transactions on Computational Science and Computational Intelligence,
DOI 10.1007/978-3-319-52419-1_1

1.2   Identifying the Need for an Interactive Approach

In a recent discussion of the growing pervasiveness and strategic importance of social media in modern societies, one of the editors of this volume (Babak Akhgar) points, along with his co-authors, to its growing ubiquity, not only in geopolitical, economic, and business spheres, but also in official responsiveness to instances of crisis and disaster. Akhgar et al. [1] helpfully define 'social media' as a collective term for those online technologies and practices employed to exchange opinions and information, promote discussion, and help to further relationships. The relevant tools therefore utilise a combination of technology, telecommunications, and accompanying social interaction. These authors follow [2] in identifying six principal varieties of social media:

* collaborative projects (e.g. Wikipedia)
* blogs and microblogs (e.g. Twitter)
* content communities (e.g. YouTube)
* social networking sites (e.g. Facebook)
* virtual game worlds (e.g. World of Warcraft)
* virtual social worlds (e.g. Second Life) (ibid.)

Akhgar and his colleagues emphasise that, in emergency situations typically characterised by the destruction and loss of landline systems (and in which the emergency services find themselves overwhelmed by the sheer volumes of calls for help or information), it is social media that invariably proves most robust, enduring, and accessible both to the general public and immediate authorities. Thus, it is undoubtedly the case that:

> Today, social media play an important role in disaster and crisis events with significant national security implications, enabling citizens' involvement through the provision, seeking and brokering of information, connecting those within and outside the event's geographical space, with implications for both the informal and the formal response effort. In this context, often characterised by underdeveloped and degraded operational environments, public safety and security organisations are required to deal with the new trend of a digitally enabled social arena in disasters and crises. Social media indeed qualify as changing the rules of the game, forging a loudly silent transformation from a need-to-know to a need-to-share principle and from a command and control to a *connect and collaborate* paradigm ([1], p. 760).

The same authors maintain that, even in the modern age where general public has become accustomed to immediacy and instantaneity, there continues to be over-reliance by majority of official PPDR (public protection and disaster relief) on traditional channels of communication, such as television and radio, despite the well-known sensationalist tendencies of such media. The messages transmitted are typically 'unidirectional' and non-interactive, and all too lacking in 'availability, details, and empathy'.

Yet, as Akhgar et al. continue to explain, social media has a proven capacity to inform, involve, and reassure the public and to assist the relevant authorities in times of emergency crisis and disaster. They point to salient examples where online social media technologies have been used to great effect by people trying to locate missing friends and relatives (e.g. in the 2010 Chilean earthquake and tsunami), to provide enlightenment and advice (as in the Chinese SARS outbreak), providing eye-witness diagnostic accounts from affected areas and thereby promoting widespread situational awareness (e.g. natural catastrophes like the 2005 Hurricane Katrina and the 2010 Haiti Earthquake, but also in the USA, London, and Madrid terrorist attacks), and providing authorities with the information required to identify the perpetrators of social atrocities as the Boston Marathon bombing.

The feeling that social media has so much to offer in situations of this nature has been resoundingly endorsed elsewhere:

> Social media can provide a window into social networks in times of crisis or disaster. The rapid reaction time of microblogging creates potential to track reactions to such events in near real time, to assess damage and suffering, and to identify individuals and networks of interest, if the relevant communications can be recognized from within the larger stream. This provides opportunities for performance improvements in emergency management, law-enforcement, and for more effective government communication and improved transparency. If such knowledge were available to everyday citizens, they might benefit from a richer understanding of information and resource flow in extreme circumstances. They might be better able to contribute to response and recovery by sharing personal knowledge or observation more effectively. Their ability to gather critical information or resources in a timely fashion could be magnified. To the degree that the social networks that emerge in times of crisis have some common core prior to the emergency, it might even be possible to proactively nurture, or plug into, these networks ([3], p. 156).

Expert commentators like Akhgar and his colleagues acknowledge that social media are not an absolute and unqualified panacea. For one thing, their 'flat and open' nature is concerning to authorities who fear that information exchanged may be deliberately or unintentionally erroneous, leading to pernicious rumour and/or misguided; used for malicious or nefarious purposes; or employed for the purpose of slander or defamation. There are related fears that such unchecked and unmonitored information may lead to the further endangerment of lives. Adding to all this are a host of practical problems, concerning such issues as interoperability (the fact that technologies used in some societies may not function properly in others) and reach (the possibility that such technologies will not be available to some sections of the population); and legal considerations relating to privacy and data protection (see also [4]).

Nevertheless, Akhgar et al. consider it essential that contemporary societies accommodate their crisis and disaster communication response systems and philosophies to an 'Information Age reality' that 'includes an informed, active and digitally empowered audience demanding authorities to present enhanced network-enabled connection and collaboration capabilities, able to handle freedom of expression, source anonymity, decentralised flows of information, sharing of information and shared situation awareness, all of which enhance effectiveness, especially in crisis environments' (op. cit., p. 764). They point to the readiness of social media providers (e.g. Google, Facebook, and Twitter) to co-operate in the push to see such channels become more established in field of crisis and disaster management. Akhgar et al. assert that relevant progress in the USA is currently well in advance of that being made in Europe, but that there is a growing determination to 'build a common crisis and emergency system'. It was in keeping with this sense of mission that the intention to develop, establish, and refine the ATHENA system was realised.

1.3   An Overview of Contents

The following outline and discussion of the ATHENA project (as an exemplar of the potential benefits arising from the use of social media and personal communications technology) is presented in four parts. The preliminary chapters appearing in Part I, *Human Factors and Recommendations for Best Practice*, are devoted, not only to accounting for apparently tentative uptake of the use of social media by European police forces and other emergency service providers during episodes of crisis management, but also to unearthing the principles of best practice pertaining to interventions involving the use of such technologies and procedures.

In their opening chapter of the book, Kerry McSeveny and David Waddington establish the important preliminary point that the strategic and tactical approaches by the police and other emergency services to crises and disasters have hitherto been predicated on perennially misguided and enduring assumptions that human behaviour in such situations is invariably panic-stricken, confused, irrational, selfish, and antisocial. Their chapter readily concurs with the views of fellow-academics, like Quarantelli [5], who maintain that it is more accurate to look upon pro-social, eminently rational and often altruistic forms of behaviour as the order of the day. McSeveny and Waddington's tentative discussion of some of the lessons of best communicative practice apparent in the literature is further developed (in Chap. 3) by Eric Stern, who offers a far more exhaustive list of strategic and practical prescriptions, predicated on a preliminary explanation of the likely impact of stress on crisis decision-making.

In their second contribution to this volume (Chap. 4), McSeveny and Waddington carefully unpack and critically analyse the nature and effectiveness of the use of social media made by first responders, emergency services, and government officials in their indicative case studies of major instances of public disorder, terrorist activity, and natural disaster.

In endeavouring to build both on the most sensible assumptions regarding human responsiveness to crisis and disaster situations, and on prescriptions of best practice, the Athena project set out to explore the ways and extent to which contemporary web-based social media and high-tech mobile devices, could be harnessed with a view to securing more efficient and effective modes of communication and enhancing the quality of situational awareness in relation to events of this nature [6]. An early statement of the project's fundamental vision maintained how:

> ATHENA will help the public help themselves by empowering them with their own collective intelligence and the means by which they can exploit that intelligence. ATHENA will provide the emergency services with new real-time intelligence from crowd-sourced information, greatly assisting in their decision-making processes and making search and rescue more efficient. ATHENA will create a fundamental and permanent shift in the way crisis situations are managed; helping the public as victim to turn into the public as part of the crisis team. ATHENA will utilise social media and smart mobile devices as part of a shared and interoperable two-way communication platform. By developing an orchestrated cycle of data, information and knowledge, ATHENA will empower both the public and emergency services with the intelligence they need in dealing with a crisis. ([6], p. 168)

It was basically assumed that, in the event of a crisis, such as a natural disaster (e.g. hurricane, forest fire, or major flood), a human agency disaster (such as a plane crash or motorway pile-up), a terrorist attack, or a large-scale public disorder involving widespread conflict and destruction of property, the public would be able to activate an ATHENA Crisis Mobile 'App', which would not only enable them to feedback reports on ongoing crisis-related activities, but also allow them to engage in any crisscrossing social media. It was further anticipated that this combination of reports and information scanned in from social media activity would be filtered into a Command and Control Centre via an intermediate information processing hub, to be used as relevant, real-time information by LEA commanders. A core part of this resource would be an ATHENA Crisis Map, highlighting such crucial information as latest updates on particular crisis- and disaster-related events, the location of emergency service personnel, of danger zones, and recommended safe routes. The ATHENA Crisis Mobile App would not only enable public users to access the crisis map and other such vital information emanating from the Command and Control Centre, but would also provide them with the means by which to appeal for help and rescue [6].

Part II of this book deals with all aspects of the Technological Design and Development of ATHENA. A useful bridge between this second part of the volume and our earlier contributions is provided by Simon Andrews' chapter (Chap. 5), which shows how the principles of best practice outlined earlier have been embodied in the design of a map-based resource for use in command and control, and an app for the benefit of the public and/or first responders. A related chapter by Kellyn Rein, Ravi Coote, Lukas Sikorski, and Ulrich Schade (Chap. 8) addresses the important need within such a system to deal effectively with the high volume and degree of complexity of the social media messages likely to be circulating during any instance of crisis and disaster. They duly set out the nature and purpose of an automated basis for helping decision-makers to cope with the multiple languages and non-relevant messages that are likely to confront them, as well as to isolate the more pertinent content of higher potential value. In Chap. 7, Chi Bahk, Lucas Baptista, Carly Winokur, Robin Colodzin, and Konstantinos Domdouzis explain how the ATHENA Mobile Application works and how it can be used to improve crisis management.

In Chap. 6, Simon Andrews, Tony Day, Konstantinos Domdouzis, Laurence Hirsch, Raluca Lefticaru, and Constantinos Orphanides elaborate on the means by which data is adapted into the form of knowledge to be used in crisis management. This discussion therefore focuses on the way in which various components of the ATHENA system engage in the 'searching, acquisition, aggregation, filtering, and presentation' of such information. This technological discussion of exactly how the system operates is taken further (in Chap. 9) by Patrick de Oude, Gregor Pavlin, Thomas Quillinan, Julij Jeraj, and Abdelhaq Abouhafe, which outlines the nature and purpose of the A-Cloud and Athena Logic Cloud, whose purpose is to facilitate the collection and presentation of data in such a way to avoid any undue or unnecessary 'information overload'.

The penultimate Part III of the volume is concerned with *Salient Legal Considerations*. Though constituting only a relatively small proportion of the entire work, the two chapters involved deal with important legal factors impinging on the ATHENA system. In Chap. 11, Fraser Sampson and Alison Lyle return to the subject of the English riots of 2011, first referred to in Chap. 4. This example is used by these authors to focus on major issues of privacy and data protection which are fundamental to the use of social media which invariably entails all sorts of information and detail of individuals relating to 'personal and sensitive aspects of their lives'. These and other human rights issues are further explored in a related chapter by Alison Lyle (Chap. 10), in which she considers the relevance to the ATHENA project of legislation and other components of the legal architecture within such bodies as the European Union and Council of Europe.

Members of the ATHENA project team have been committed from the outset to the rigorous and ongoing evaluation, not only of the component technical features, but also of the platform as a whole. This strong evaluative ethos has been reflected in the occurrence of four 'test bed' exercises, carried out at appropriate stages of the project's lifetime. This approach has involved a series of case studies in which lifelike scenarios involving public disorder, terrorism, and natural disaster have been used as a basis for testing and subsequently improving the effectiveness of the core elements of the ATHENA system and, more latterly, the degree to which they are capable of successfully integrating.

The final two chapters comprising Part IV of this volume, *Testing the ATHENA System*, refer to the nature and outcomes of this process of evaluation. Chapter 12 thus focuses on the findings and implications of three preliminary case studies (one in Izmir, Turkey; and two more in Ljublana, Slovenia), concerned with the ongoing development and iteration of the relevant technology. Chapter 13 then goes one step further by reporting on the nature and outcome of a full-scale examination of the fully customised system, held in the West Yorkshire city of Wakefield in the United Kingdom.

Thus, we are left with the concluding chapter of this book, which amounts, as one might expect, to a closely considered final reflection on the overall evolution of the ATHENA project: the extent to which it can justifiably claim to have achieved its objectives; its overall effectiveness, and the extent of any obvious need and possible methods of improvement. This conclusion will also re-emphasise the point that, whilst we have been focusing here on a single, concrete example, it has been our wider intention to show how the underlying human, social, legal, and technological considerations pertaining to our work have the clearest possible resonance to any future initiatives of this nature.

References

1. Akhgar, B., Fortune, D., Hayes, R., Guerra, B., & Manso, M. (2013). Social media in crisis events: Open networks and collaboration supporting disaster response and recovery. In *Technologies for Homeland Security (HST), 2013 IEEE International Conference*. IEEE, 2013 (pp. 760–765).
2. Kaplan, A. M., & Haenlein, M. (2010). Users of the world, unite! The challenges and opportunities of Social Media. *Business Horizons, 53*(1), 59–68.
3. Glasgow, K., & Fink, C. (2013). From push brooms to prayer books: Social media and social networks during the London riots. In *iConference 2013 Proceedings*. February 12-15, 2013. Fort Worth, TX, USA (pp. 155–169).
4. Lindsay, B. (2011). *Social media and disasters: Current uses, future options, and policy considerations*. Congressional Research Service: Washington, DC.
5. Quarantelli, E. L. (1993). Community crises: An exploratory comparison of the characteristics and consequences of disasters and riots. *Journal of Contingencies & Crisis Management, 1*(2), 67–78.
6. Andrews, S., Yates, S., Akhgar, B., & Fortune, D. (2013). The ATHENA Project: using Formal Concept Analysis to facilitate the actions of respondents in a crisis situation. In B. Akhgar & S. Yates (Eds.), *Strategic intelligence management*. London: Elsevier.

Part I
Human Factors and Recommendations for Best Practice

Chapter 2
Human Factors in Crisis, Disaster and Emergency: Some Policy Implications and Lessons of Effective Communication

Kerry McSeveny and David Waddington

2.1 Prevalent Myths and Misconceptions

It is a commonly held misconception within lay society and also amongst relevant official authorities that people are apt to respond to acute situations of crisis, disaster and emergency in a socially disorganised and individually disoriented fashion. As Perry and Lindell explain,

> Decades of 'disaster' movies and novels and press coverage, emphasise the general theme that a few 'exceptional' individuals lead the masses of frightened and passive victims to safety. Thus, conventional wisdom holds that typical patterns of citizen disaster response take the form of panic, shock, or passivity. ([1], pp. 49–50)

Drury et al. [2] have elaborated on this perspective by highlighting three contemporary myths that contribute to a misunderstanding of civilian cognitions and emotions in situations of this nature. Firstly, the myth of 'mass panic' wrongly presupposes that people typically respond to the exaggerated, 'contagious' and irrational fears that inevitably engulf them by engaging in overhasty and ill-advised escape behaviours which seem unrestrained by any recognisable social rule or convention. The second myth of 'helplessness' is predicated on the equally misguided assumption that people immediately become too stunned or 'frozen' to adequately ensure their own safety and well-being. Finally, the 'civil disorder' myth is based on the unfounded notion that emergency situations provide a context or 'excuse' for people to behave in antisocial and/or opportunistic behaviours, such as rioting and looting.

All of the above authors point to varied and compelling evidence in rebuttal of misconceptions of this nature.

K. McSeveny (✉) • D. Waddington
CENTRIC, Sheffield Hallam University, Sheffield, UK
e-mail: K.McSeveny@shu.ac.uk; D.P.Waddington@shu.ac.uk

© Springer International Publishing AG 2017
B. Akhgar et al. (eds.), *Application of Social Media in Crisis Management*,
Transactions on Computational Science and Computational Intelligence,
DOI 10.1007/978-3-319-52419-1_2

Indeed, most citizens do not develop shock reactions, panic flight occurs only rarely and people tend to act in what they believe is their best interest, given their limited understanding of the situation. Most citizens respond constructively to environmental threats by bringing as much information and as many resources as they can to bear on the problem of how to cope with an incident. Behaviour in the disaster response period is generally prosocial as well as rational. Following impact, uninjured victims are often the first to search for survivors, care for those who are injured, and assist others in protecting property from further damage. Antisocial behaviour such as looting is relatively rare, while crime rates tend to decline following disaster impact ([1], p. 50).

In attempting to account for such essentially prosocial and often altruistic behaviour, theorists have leaned towards the so-called affiliation model and/or 'normative' approaches to explaining emergency or disaster-related behaviour [3]. The former posits that, rather than primarily 'looking after Number One', individuals tend to prioritise the safety and security of those people who are (biologically, socially or emotionally) closely related to them (ibid.). The latter rests on the equally straightforward and simple assumption that civilian behaviour in emergencies generally adheres to an equivalent set of rules to those governing everyday social conduct (ibid.). Thus, observations of people helping others to evacuate from buildings on fire reveal that most help is accorded to such 'vulnerable' people as the elderly, and that customary chivalry tends to endure to the extent that men are especially supportive of women (ibid.).

There are, however, fundamental problems with each of these explanations. In the first place, while the affiliation model might well help to explain the prosocial behaviours occurring in situations involving family or friends, most emergencies and disasters tend to involve aggregations of complete strangers, having no previous personal ties. Second, it is also the case that, 'While it might be normative to help someone in distress in everyday circumstances, it is surely novel rather than normative to take risks to oneself help strangers' (ibid., p. 10). Thus, it is clear that what is required in order to complement and overcome the limitations of these approaches is a 'model of mass emergent sociality' (ibid., p. 11, emphasis in original).

2.2   Shared Fate and Unity of Purpose

Drury and Cocking [3] have utilised a variety of experimental and ethnographic approaches to address this very requirement. Their interviews with 21 survivors from a variety of emergency scenarios (e.g. sinking ships, bombings and football stadium disasters) highlight in particular that the sense of shared fate experienced by those involved tends to induce a powerful shared sense of identity and unity of purpose:

In most of the references to common identity, it is described as emerging over the course of the emergency itself. Only a minority referred to any sense of crowd unity prior to or without there being a perceived emergency—and for most of these the sense of unity increased in response to the emergency. The source of the unity was the crowd members' shared fate in relation to the threat facing them. While they might have come to the event seeing themselves as so many individuals, the threat facing them all led them to see themselves as 'all in the same boat'. (ibid., p. 20)

These authors emphasise that the disaster and emergency situations they observed (often characterised by conditions of extreme danger and the possibility of death) were 'occasions for the display of the noblest intentions and behaviours rather than the basest instincts' (ibid., p. 29).

Elsewhere, Drury and/or Cocking have used numerous case studies to illustrate the fact that people's behaviour in such circumstances was invariably 'orderly and meaningful', with signs of selfish or uncooperative behaviour being few and far between. Those rare instances of outright 'selfishness' that did occur were never imitated by others in the vicinity, and the individuals in question often found themselves being sternly rebuked by those around them ([4], p. 69). Cocking [5] likewise relates that, while people undoubtedly looked around, in the initial absence of the emergency services, for some sort of direction, they were invariably discerning and by no means uncritical of attempts to exercise leadership and influence.

This was evident in the immediate reaction of individuals caught up in the London Underground bombings of 7 July 2005 ('7/7'). During the 45-min period before the emergency providers arrived, those present on one of the trains affected were far more responsive to the 'calmer', more reassuring form of leadership spontaneously exhibited by a female solicitor, than to an allegedly 'stupid' man, who was 'too full of his own importance'. In the words of one actual eyewitness:

> I think people seemed to be glad that there was somebody like the lawyer woman taking some kind of control [...] I think people looked to that [...] and she had a good strong voice, she was sensible, she commanded some kind of respect and authority if you like and what she was saying was very sensible so people were taking note [...] the bloke he was just a bit of a pompous ass and I don't think people were really taking much notice of him. (quoted by [5], p. 88).

Cocking and Drury [6] echo the conclusions of researchers like Cole et al. [7] who maintain that the compassionate tone exuded by informal leadership of this nature is often in stark contrast to the somewhat brusque 'command-and-control' ethos exhibited by emergency services, most notably the police. During the Hillsborough stadium disaster of April 1989, for example, football fans finding themselves trapped in massively overpopulated spectator enclosures helped one another to escape the confines imposed by 2-m security fences. Indeed,

> Despite the predominant image of football fans at the time as violent hooligans, the crowd's response was quiet and considered, with individuals assisting one another and calming the situation down...The public assisted one another and carried the injured to the ambulances outside the stadium, preventing a potentially higher death toll. ([7], pp. 367–8)

By contrast, senior and junior South Yorkshire Police officers alike seemed to regard the matter as 'a public order, rather than public safety issue' ([5, 8], p. 80)—so much so that one eyewitness was allegedly told to '[f....off]' on appealing to a junior police officer to throw open a nearby gate and generally make more effort to organise the crowd [6].

This certainly chimes with related research on the policing of public disorder, which highlights a corresponding tendency for pervasively held myths and misconceptions about the dispositions and dynamics of crowds of protesters to produce ill-conceived and counterproductive policing interventions. Thus, as Drury et al. [2] point out in their essay on disaster myths,

> Convergent evidence for this line of argument comes from research on 'public order' policing. Pathologizing representations of the mass (e.g., the 'mad mob') have been shown to rationalize coercive policing practices....which offend the peaceful crowd's sense of legitimacy and in turn produce the very angry, 'disorderly mob' that the police presumed.

The Elaborated Social Identity Model (ESIM) presupposes that police crowd control interventions can often serve to aggravate potential conflict to the extent that they appear unreasonable and/or indiscriminate to those present. Such tactics may well have the inadvertent effect of instilling amongst crowd members a perception of their shared fate and identity, and feelings of solidarity in the face of a common foe. Thus, even those participants harbouring no prior intention of engaging in confrontation with the police may be drawn into the ensuing conflict: 'We find that people who expect the police to uphold their democratic rights (to protest, to watch sport in safety) but feel that the police have denied these rights are often those who are most outraged, most angry and who enter the subsequent crowd events with the greatest willingness to confront the police' ([9]: 564)—See also Chap. 3.

This phenomenon has been more recently examined by Cocking [8], who conducted in-depth interviews with 20 respondents who had first-hand experience of having been subjected to police dispersal charges. Cocking reports that, although such respondents confessed to an initial feeling of fear on having been charged at by the police, this soon subsided to be replaced by a growing feeling of determination and sense of unity amongst the crowd, 'suggesting that a shared sense of collective identity had emerged from the initially fearful experience' (ibid., p. 226). Thus, far from physically and psychologically fragmenting the crowd (in keeping with the objectives of the exercise), police dispersal tactics actually produce a unifying effect which enhances the prospect and intensity of disorder.

2.3   Policy Implications

There is a growing consensus amongst social scientists, managers of responses to emergencies and disasters, and humanitarian agencies that myths of this nature do have considerable policy implications: 'The myths of irrational and antisocial behaviour in disaster are not just erroneous—they hamper the effectiveness of emergency planning by misdirecting the allocation of resources and the dissemination of information' ([1], p. 50). Thus, as Drury et al. [2] point out, both 'panic' and 'helplessness' myths are known to underlie the restriction or withholding of information by the authorities in relation to various political or environmental threats. Perry and Lindell add that, 'This response to the myth of panic is particularly troubling because it has been shown repeatedly that people are more reluctant to comply with suggested emergency measures when they are provided with vague or incomplete information (warning messages)' ([1], p. 50).

An ill-conceived adherence to the 'public order' myth can also be decidedly unhelpful: for example, the unfounded expectation that Hurricane Katrina was bound to result in an upsurge of opportunistic looting led to a military rather than humanitarian reaction by the American authorities ([2] op. cit., p. 2260). As Auf der Heide [10] maintains, fears of this nature can prove dysfunctional in several important ways:

> For example, one reason people refuse to evacuate in disasters is to protect their property... It is also ironic that security measures undertaken to 'prevent looting' can prevent residents from salvaging property that is exposed to the elements by the disaster...Finally, overzealous police and security guards manning roadblocks set up to keep looters out sometimes prevent the entry of legitimate disaster-response personnel.

Perry and Lindell [1] emphasise that it is important not to confuse the type of fear and anxiety that may reasonably be expected in situations of crisis, emergency and disaster with panic-stricken or senseless behaviour. Given that people's knowledge about such vital issues as (say) the chemical, biological or radiological agents used in a terrorist attack is bound to be extremely limited, it is important for the authorities to urgently disseminate relevant information regarding the possible hazards involved alongside recommended means of protection:

> One need not try to give those at risk a broad education about these topics, just specific relevant information. Officials should focus on defining the threat, explaining its human consequences, and explaining what can be done to minimise negative consequences. If the actions to minimise the consequences cannot be undertaken by individuals, but must be executed by authorities, then one explains what is being done. Contrary to popular fiction, the road to anxiety reduction is through providing—not withholding—information. (ibid., p. 54)

These authors make the reassuring point that situations involving the presence of an unfamiliar threat generate circumstances in which citizens automatically look to the authorities for guidance, and in which both their attention to messages from the emergency agencies and readiness to take heed of official recommendations are generally at their height. This makes it imperative, of course, for all communication between the different agencies involved to be as closely coordinated as possible, such that each agency is totally aware of the nature and limitations of their own role, and the corresponding functions and responsibilities of those occupying related roles (ibid.)

The way in which the authorities might choose to relate to relevant sections of the public is also a matter of great significance. While the withholding of information can lead to a lack of trust in the authorities, it also signifies that those authorities lack trust in the public to react in purposeful and useful ways in an emergency situation. Drury and Cocking note the 'resilience' of the crowd, and suggest that 'the ability of the crowd to provide mutual aid, to co-ordinate and co-operate, to deal with individual distress and panic, to take initiatives and play a leadership role should not be underestimated' (op. cit., p. 32).

Recent research on civilian responses to emergency and disaster situations suggests that the authorities should see the crowd as 'part of the solution, not the problem' [7], treating the public as a 'capable partner' ([11], p. 218). In fact, failure to acknowledge the potential benefits which could arise from the public's willingness to help in such conditions means that a vast potential resource is currently being ignored. Drury and Cocking accept that public attempts at 'helping' are not always actually helpful in reality, but they nonetheless insist that 'the blanket exclusion of the public from emergency planning, and the treating of crowd members simply as victims, may be counter-productive' (ibid., p. 32).

Treating the public as victims can also have further implications, especially as 'the over-protective responses of the government may stunt the public's own natural resilience' (Drury and Cocking, [3], p. 32). This also informs the way in which any messages should be formulated. Veil et al. [17] recommend that any communication should include messages of self-efficacy, an approach which not only provides members of the public with a sense of control, but also clearly outlines what individuals can do to help.

The concerns of the public should also be taken into account, both during and before an event, and Veil et al. [17] suggest establishing a dialogue between the authorities and the public, since 'listening to and understanding a public through monitoring public opinion about risk is essential in the development of a relationship' (p. 111). Glass and Schoch-Spana [11] note the importance of participatory decision-making processes, and it is advised that other stakeholders should also be involved in this dialogue—which ought to be 'aimed at resolving disputes and reaching consensus' [12].

2.4   Lessons for Communication Strategy

Drury and Cocking [3] point out that, given the importance of effective two-way communication, priority should be given to the creation and maintenance of reliable communication channels (such as public address systems) rather than physical features of disaster or emergency prone public places (such as the width of exits). They also make several recommendations regarding the possible media of communication that are liable to prove most effective. Thus, they advocate the use of public announcement systems rather than sirens, and suggest that greater utility is made of video screens.

Channels of communication must also be reconsidered and reappraised in terms of their effectiveness. Sellnow and Vidoloff ([13], cited in [14]) stress that it is extremely important that any communication which takes place in a crisis situation to be suitably sensitive to any cultural differences between groups affected by the situation. Thus, messages should be disseminated in ways which 'make additional efforts to reach under-represented populations including those who are enduring poverty, are new to [the country], or others who have limited access to mainstream media' ([14], p. 112).

There is an equally obvious need for the source of any communication with the public to be perceived as both credible and trustworthy. It is almost inevitable that the role of 'leader' will also be informally taken on by one or more members of the crowd, with a view to helping to calm other victims, or create a return to order (as happened, for example, in the '7/7' bombings in London). However, due to the need in such situations for accurate and consistent information [14], it is advisable to ensure that there is a rapid deployment at ground level of 'wardens', such as fire brigade personnel, police officers or stewards, to whom people can look for guidance and direction.

These individuals must be familiar with the area, and sufficiently trained to be able to provide consistent and confident information to the public. This ties in with Kapucu's [15] notion of 'boundary spanners'—i.e. those organization members 'who link their organisation with the external environment'. Kapucu notes that it is vital to have people in place who are capable of understanding the composition, cultures and sensibilities of the various group or groups involved, and thus able to decide the best methods of effectively sharing and disseminating relevant information across group and organisational boundaries.

Trust may also be related to how much the public/crowd relate to the source in question—for example, they may find it easier to identify with, and thus have confidence in, a local radio station, rather than a government spokesperson. Relationships with credible sources can be developed before a crisis occurs [17].

Ritchie et al. [14] stress the importance of a quick response to an event, ensuring that relevant information is communicated to stakeholders as soon as possible. The instructions provided during emergencies also need to be 'clear, informative and easily accessible to the public', providing 'the knowledge needed for informed decision making about risks' ([12], p. 383). Evidence suggests that access to information increases an individual's chance of survival in an emergency situation, but this is not always made as readily available as it could be (e.g. tube train information during the 7/7 London bombings).

The key role played by information dissemination in an evacuation context means that, as noted previously, the communication should be 'explicit and unambiguous' to ensure it is as effective as possible. This information should also be accurate and consistent, and should inform stakeholders what has happened (where, when and how), what is being done about the situation and what precautions they can take [14]. Hesloot and Ruitenberg ([16], p. 105) suggest that the focus of any efforts should be geared towards 'solving the problem, not the prevention of chaos'. However, despite the need for clear and accurate communication, Veil et al. [17] note that emergency situations are often inherently ambiguous, and so a certain level of uncertainty must be accepted in communications, which can be modified as more information becomes available.

Perry and Lindell add that information of this nature should extend to the recommendation of particular courses of action. They maintain that, once informed of the presence of a particular form of hazard, individuals will naturally try to undertake any steps they believe necessary to reduce the danger (See Chaps. 3 and 5). Thus,

A message not accompanied by constructive suggestions for action simply enhances fear, which itself cannot be salved without information and action. When providing protective action recommendations, it is also critical to briefly link the action with protection for the citizen. Telling citizens why evacuating an area will reduce their exposure to smallpox, or why taking potassium iodide will reduce radiation exposure damage accomplishes two important objectives. First, it increases compliance by those at risk, and second, it discourages them from taking other actions that seem to be effective but are not [1], pp. 54–55.

The tone of any communication is also vitally important when attempting to encourage the public to trust in any official authority or spokesperson. Veil et al. [17] suggest that spokespersons should humanise the situation as much as possible and demonstrate genuine commitment by communicating with 'compassion, concern and empathy'. The potential resilience of the crowd is directly related to their shared identity. In other words, the more collective spirit is fostered, the more the crowd is likely to be to respond effectively to an emergency situation: This has important consequences for the way in which information is communicated to the public. Messages which appeal to the collective spirit of the crowd, rather than referring to them as a group of isolated individuals, are more likely to foster a sense shared identity and thereby encourage cooperation.

2.5 A Place for Social Media

Historically at least, the police in western societies have been somewhat sluggish in their uptake and utilisation of social media as a means of engaging with and encouraging the general population at large (e.g. [18, 19]. One notable exception investigated by the present authors [20, 21] was the social media strategy devised and implemented by South Yorkshire Police in response to the staged protest by 5000 people occurring outside the Liberal Democrats' Spring Conference in Sheffield in March 2011. An internal memorandum issued in January 2011 defined the main aims and objectives of the strategy as: maintaining public confidence; engaging with social media communities and potential protesters; and providing 'consistent and informed messages' to the parties concerned [20]. The strategy involved the use of Twitter, Facebook, and local and national discussion forums, both in the build up to and during the event (ibid).

At the pre-conference stage, the Senior Media Officer in charge of the 4-person Social Media Cell assigned to the operation worked for several months to build up as significant a Twitter following as possible. Tweeting in specific relation to the event started in earnest a month prior to the protest, with the SMO taking personal responsibility for maintaining the account—which was attributed to her as an individual, rather than corporately to 'South Yorkshire Police'. At this point in time, regular updates were posted, using the hashtag #libdempolicing. Such messages were deliberately upbeat and positive in nature. For example:

@InspJForrest: SYP are committed to providing a safe and enjoyable environment for all through their #libdempolicing. More details to follow (quoted in ibid).

On the day of the actual conference, Twitter continued to be used extensively, not only in order to maintain an upbeat ethos, but also to rectify any misunderstandings related to particular police tactics. The SMO moved swiftly, for example, to repudiate assertions that a gated barrier at the bottom of the protest site was being used with the intention of 'kettling' protesters. She quickly succeeded in persuading her Twitter followers that the gate was being used purely as a safety measure—to offset any possibility of crushing. An accompanying reassurance was consistently put out to the effect that 'South Yorkshire Police does not acknowledge kettling as an approved Home Office method of crowd control' (ibid.). Tweets were also successfully utilised in order to scotch potentially pernicious rumours—such as a rapidly growing theory circulating amongst the crowd that police personnel on top of the John Lewis department store, directly opposite the protest site, were armed snipers in disguise.

The police's own impression that its social media strategy contributed to an outcome involving only one arrest is well borne out by the substantial positive feedback subsequently volunteered by members of the public. The SMO appeared justified in attributing this success to the fact that she was allowed to make her Twitter posts in her capacity as an individual, as opposed to a corporate account. In this way, respondents were encouraged to engage with her 'on quite a personable level' (ibid.).

While undoubtedly highlighting the strategic and tactical utility of the police use of social media in uncertain and socially volatile situations, the impact and significance of our example was massively superseded by the occurrence of the English riots later that same year (see, for example, [22]). The widespread social mayhem, conflict and physical destruction characterising these riots certainly qualifies them for inclusion amongst the three case studies of crisis situations or events we intend using in Chap. 4 of this edition to investigate the way that social media has been employed by the police and emergency services in response to large-scale public disorder, terrorism and natural disaster and global disease (See Chap. 6). The case studies in question are presented as a basis for exploring in close detail some lessons of good and bad practice, and the key benefits and risks, associated with incorporating social media and personal communication technology capabilities into crisis preparedness and management.

References

1. Perry, R., & Lindell, M. (2003). Understanding citizen response to disasters with implications for terrorism. *Journal of Contingencies & Crisis Management, 11*(2), 49–60.
2. Drury, J., Novelli, D., & Stott, C. (2013). Psychological disaster myths in the perception and management of mass emergencies. *Journal of Applied Social Psychology, 43*, 2259–2270.
3. Drury, J., & Cocking, C. (2007). *The mass psychology of disasters and emergency evacuations: A research report and implications for practice.* Brighton: University of Sussex.
4. Cocking, C., Drury, J., & Reicher, S. (2009). The psychology of crowd behaviour in emergency evacuations: Results from two interview studies and implications for the Fire and Rescue Services. *The Irish Journal of Psychology, 30*(1–2), 59–73.

5. Cocking, C. (2013). Crowd flight in response to police dispersal techniques: A momentary lapse of reason? *Journal of Investigative Psychology and Offender Profiling, 10*(2), 219–236.
6. Cocking, C., & Drury, J. (2014). Talking about Hillsborough: 'Panic' as discourse in survivors' accounts of the 1989 football stadium disaster. *Journal of Community and Applied Social Psychology, 24*, 86–99.
7. Cole, J., Walters, M., & Lynch, M. (2011). Part of the solution, not the problem: The crowd's role in emergency response. *Contemporary Social Science: Journal of the Academy of Social Sciences, 6*(3), 361–375.
8. Cocking, C. (2013). The role of 'zero responders' during 7/7: Implications for the emergency services. *International Journal of Emergency Services, 2*(2), 79–93.
9. Reicher, S. D., Stott, C., Cronin, P., & Adang, O. (2004). An integrated approach to crowd psychology and public order policing. *Policing: An International Journal of Police Strategies & Management, 17*(4), 558–572.
10. Auf der Heide, E. (2004) Common misperceptions about disasters: panic, the "disaster syndrome" and looting. In M. O'Leary (Ed.), *The First 72 Hours: A Community Approach to Disaster Preparedness*. Lincoln, Nebraska: iUniverse Publishing.
11. Glass, T., & Schoch-Spana, M. (2002). Bioterrorism and the people: How to vaccinate a city against panic. *Confronting Biological Weapons, 34*, 217–223.
12. Covello, V., Peters, R., Wojtecki, J., & Hyde, R. (2001). Risk communication, the West Nile virus epidemic, and bioterrorism: Responding to the communication challenges posed by the intentional or unintentional release of a pathogen in an urban setting. *Journal of Urban Health: Bulletin of the New York Academy of Medicine, 78*(2), 382–391.
13. Sellnow, T., & Vidoloff, K. (2009). Getting crisis communication right. *Food Technology, 63*(9), 40–45.
14. Ritchie, B., Dorrell, H., Miller, D., & Miller, G. (2004). Crisis communication and recovery for the tourism industry. *Journal of Travel & Tourism Marketing, 15*(2–3), 199–216.
15. Kapucu, N. (2006). Interagency communication networks during emergencies: Boundary spanners in multiagency coordination. *The American Review of Public Administration, 36*, 207–225.
16. Helsloot, I., & Ruitenberg, A. (2004). Citizen response to disasters: A survey of literature and some practical implications. *Journal of Contingencies & Crisis Management, 12*(3), 98–111.
17. Veil, S., Buehner, T., & Palenchar, M. (2011). A work-in-process literature review: Incorporating social media in risk and crisis communication. *Journal of Contingencies & Crisis Management, 19*(2), 110–122.
18. Brainard, L., & McNutt, J. (2010). Virtual government-citizen relations: Informational, transactional or collaborative? *Administration and Society, 42*, 836–858.
19. Crump, J. (2011) What are the police doing on Twitter? Social media, the police and the public. *Policy and Internet, 3*(4), article 7.
20. McSeveny, K., & Waddington, D. P. (2011). Up close and personal: The interplay between information technology and human agency in the policing of the 2011 Sheffield Anti-Lib Dem protest. In B. Akghar & S. Yates (Eds.), *Intelligence management (knowledge driven frameworks for combating terrorism and organised crime)*. New York: Springer.
21. Waddington, D. (2013). A 'kinder blue': Analysing the police management of the Sheffield anti-Lib Dem protest of March 2011. *Policing and Society, 23*(1), 46–64.
22. Moran, M., & Waddington, D. P. (2016). *Riots: An international comparison*. London: Palgrave Macmillan.

Chapter 3
Crisis Management, Social Media, and Smart Devices

Eric K. Stern

3.1 Introduction

It is easy to get the impression that contemporary society is more crisis prone than ever. European capitals such as Brussels, Paris, Ankara, Oslo, Madrid, and London—like US cities such as New York, Washington DC, and Boston, Orlando, US —have been the targets of dramatic terrorist attacks over the last 20 years. Volcanic ash has brought European air transport to a standstill. There have been major outbreaks of foodborne illness (e.g., 'Mad Cow' disease and EHEC) and severe scares regarding pandemic influenza, SARS, MERS, and Ebola across Europe and the globe. Natural disasters—forest fires, windstorms, flooding, earthquakes, avalanches—have wrought havoc in Europe from Turkey to Iceland and across the Americas from Chile to Canada. Increasingly, these dramatic public events are experienced, documented, communicated, and understood via social media and smart devices (cell phones, tablets, computers) by participants—citizens, official responders, leaders—and observers alike.

Though the language we use to describe them and the information and communication technologies used to document and communicate about them have evolved over the millennia, communities—and those who lead them—have always faced such threats to public safety, welfare, and order. Stone age societies and their leaders were (and in some remote corners of the world probably still are) confronted by existential threats stemming from human antagonists, infectious diseases, disturbances in the food supply, and climatic extremes just like their contemporary counterparts. While the core challenges of coping with such threats are enduring in many respects, our communities and the broader societies in which they are embedded

E.K. Stern (✉)
CRiSMART, Swedish Defense University, Stockholm, Sweden

University at Albany, SUNY, Albany, NY, USA

© Springer International Publishing AG 2017
B. Akhgar et al. (eds.), *Application of Social Media in Crisis Management*,
Transactions on Computational Science and Computational Intelligence,
DOI 10.1007/978-3-319-52419-1_3

have grown dramatically in size, scope, and complexity. Furthermore, the socio-technical context of what we now call crisis management has changed dramatically, not least in recent decades.

In other words, though crisis management is clearly a very old game, the playing field has changed in important ways. Though there are certainly national and regional differences, the following are just a few examples of trends that profoundly affect conditions in European and other liberal democratic societies around the world [1–4]:

- Globalization: Threats—and potentially effective responses to them—increasingly transcend national boundaries. This is true not only of scourges like terrorism and infectious diseases, but also of financial turbulence, environmental change, and disruption of critical infrastructure among many others.
- Multi-level, multi-sectoral governance: Crisis issues—such as those mentioned above—are managed and impacted by multiple levels of governance from the international to the local. For example, combatting infectious diseases and other threats to public health involves international organizations/arrangements such as the World Health Organization and the International Health Regulations, regional organizations such as the European Center for Disease Prevention and Control, as well as a wide range of national, regional, and local counterparts. Furthermore, not only international and domestic government organizations are involved, but also a variety of private sector (e.g., pharmaceutical companies) nonprofit organizations (e.g., Doctors Without Borders and the Red Cross).
- Politicization: Strong taboos have traditionally constrained the exploitation of crises and community tragedies for political purposes in many Western democracies, and pressures to rally around the flag and suspend partisan differences (at least during the acute phases of crises) have remained strong. These taboos and norms have been greatly weakened in recent decades and even the acute phase of crisis management has become increasingly politicized.
- Mediatization: Changes in media structure and norms in many countries have also changed the crisis management playing field in important ways. The rise of the 24 h news cycle, media internationalization, proliferation, and increased competition among global and local media actors, web-based media, and media convergence (in which traditional print and broadcast media become intertwined with web-based media and each other), evolving forms and norms of both investigative and sensational journalism, and the rise of social media and smart devices (more on this below) have profound implications for crisis management.
- Technology: Technological development is a double-edged sword which profoundly effects both the nature of the threats and risks facing contemporary societies and the prospects for coping with them. Technology enables the large-scale critical infrastructural systems that support communities (more on this below) though dependence on these systems can create cascading crises when they fail or are intentionally disrupted through physical or cyber-attacks. Technology can be used to protect the public from biological, chemical, and nuclear threats, but also can be used to create weapons of mass destruction. Information and

communications technology can be used to enable effective public and community responses [5], but also as an instrument for recruiting and radicalizing potential terrorists and orchestrating attacks (See Chaps. 2, 5 and 6).

Having set the stage a bit in the preceding paragraphs, the rest of this chapter will explore the following questions:

- What is a crisis and how can a rigorous conceptualization of crisis help crisis managers (both official and citizen responders) to make sense of potential crises?
- What are some of the key and recurring leadership tasks associated with crisis management and how are they affected by the rise of social media and smart devices?

The chapter will conclude with a number of prescriptive reflections for those engaged in the work of improving crisis management capacity for governments and communities.

3.2   Defining Crisis and Identifying Key Questions for Crisis Sense-Making[1]

As noted earlier in this volume, the point of departure for project ATHENA is five major usage cases, exemplifying different types of contingencies facing governments and communities (See Chaps. 5, 12, and 13). These include, but are not limited to:

- Terrorist incident
- Airline disaster
- Public disorder incident
- Natural disasters
- Acute threats to public health stemming from outbreaks of infectious disease/pandemics

These should be considered examples of contingencies threatening many European communities on an ongoing basis. Furthermore, because these contingencies involve somewhat different sets of government and societal/community actors and challenges, designing the ATHENA system around them has contributed to making ATHENA solutions compatible with "all hazards" approach (as opposed to a system specifically tailored and usable only for one type of contingency).

Though these contingencies have their specific features (and problem sets), they also share common qualities as potentially acute problems for members of the community, private, and nonprofit sector organizations, governments, the EU, and international organizations (see Chaps. 4, 5, and 7).

---

[1] This section is adapted from contributions by Eric Stern to ATHENA report 3.1 [6].

For the present purposes, crisis is conceptualized in terms of three subjective criteria experienced by crisis decision-makers, be they citizens, first responders, or strategic leaders: threat, uncertainty, and urgency ([7, 8]; cf. [9]). These dimensions are helpful in distinguishing crises from other types of situations but also provide a means for probing and preparing to act in crises.

1. First, crises are associated with threats to (and on occasion potential opportunities to promote) core values cherished by communities and their leaders. These values include human life, public health and welfare, democracy, civil liberties and rule of law, economic viability, functioning of critical infrastructure, and public confidence in leaders and institutions. Crisis copers must also be prepared to cope with conflicts among such values [10].
2. Second, crises exhibit high degrees of uncertainty regarding the nature of such threats (i.e., the known and unknown unknowns), the contours of an appropriate response, and the possible ramifications of various courses of action.
3. Third, crises are associated with a sense of urgency. Events are perceived as moving quickly, and there are fleeting windows of opportunity to influence their course. Additional pressure can stem from the relentless pace of the 24-h news cycle. Decision-makers and their organizations must cultivate the capacity to diagnose situations and formulate responses under severe time constraints. Thus, crises force decision-makers to make some of the most consequential decisions in public life under extremely trying and potentially stressful circumstances.

Situations in which these three conditions coincide tend to be experienced as highly stressful. As a result, the ATHENA project team has been mindful that ATHENA solutions are intended to be used under relatively extreme conditions and by users who may be subjected to high levels of stress.

Excessive levels of stress are thought to impact heavily upon individual and collective information processing. For example, a wide range of specific effects have been identified by researchers working with experimental, historical, and field studies. For example, under heavy stress, individuals have been found to:

• focus on the short term to the neglect of longer term considerations
• to fall back on and rigidly cling to longstanding behavior patterns (often forgetting more recent ones)
• narrow and deepen their span of attention, scrutinizing 'central' issues while neglecting 'peripheral' ones
• and to be prone to irritability ([11, 12]: 25–37; [13, 14]: 97–139; [15])

Some researchers argue that the laboratory-based stress literature has missed important aspects of the relationship between stress and performance errors in real world problem-solving under adverse conditions. For example, organizational psychologist Gary Klein has argued that the more important effects of stressors such as extreme time pressure, noise, and ambiguity are that:

• Stressors interfere with information gathering
• Stressors disrupt the ability to recall necessary information
• Stressors distract decision-makers' attention from critical tasks ([16]: 275).

Clearly, all of these effects can impair the ability of decision-makers to produce adequate situational assessments and make sense of crisis situations. Thus, the literature on the relationship between stress and decision-making, while not entirely pessimistic regarding the possibility of effective coping, tends to emphasize the many ways that situational assessments can be distorted by stress effects (see Chap. 13).

From the perspective of ATHENA, developing functionality to help leverage individual and collective capacity and facilitate coping with crisis conditions is a central focus. In other words, the applications and tools under development should help users to make better sense of, communicate about, and enable more effective individual and collective response to crisis situations (see the following section for a more general discussion of sense-making in crisis).

Confronted with a threatening situation, it is useful to turn the components of this crisis definition into diagnostic questions [17].

- What core values are at stake in this situation? This question helps crisis managers identify threats and opportunities embedded in the contingency at hand and encourages them to craft solutions that attend to those threats and opportunities in a consciously balanced and measured way (cf. [18]). It also helps them minimize the risk of the so-called type 3 error: the deploying the 'right' solution to the wrong problem [19].
- What are the key uncertainties associated with the situation, and how can we reduce them? This question enables decision-makers and others confronted with crisis to identify key variables and parameters and better prioritize 'intelligence' and cognitive/analytical resources.
- How much time do we have? Effective and legitimate crisis decision-making and communication processes may look very different indeed depending on whether the time frame is measured in minutes, hours, days, weeks, or months [20].

3.3   Unpacking Crisis Leadership Tasks and the Implications of Social Media and 'Smart' Devices

Several decades of intensive empirical research and practical experience of crisis management in contemporary governmental/nongovernmental settings show that organizations and their leaders face recurring challenges when confronted with (the prospect of) community, societal, or international crises [1]. These are sense-making, decision-making and coordination, meaning-making, accounting, and learning (see Chap. 7 for example of Mobile Apps).

As noted above, sense-making in crisis refers to the challenging task of developing an adequate interpretation of what are often complex, dynamic, and ambiguous situations (c.f. [21, 22]). This entails developing not only a picture of what is happening but also an understanding of the implications of the situation from one's own vantage point and that of other salient stakeholders. As Alberts and Hayes [23] put it: 'Sensemaking is much more than sharing information and identifying patterns. It goes beyond what is happening and what may happen to what can be done about it.' Prior to a crisis, sensemaking is difficult due to attention scarcity, weak or conflicting signals regarding mounting threats, and a high degree of uncertainty. Once it is clear that a crisis has occurred, a paradoxical combination of information overload and lingering uncertainty/scarcity regarding key parameters is common. Here, a number of distinctive and highly important processes become relevant.

Decision-making and Coordination refers to the fact that crises tend to be experienced by crisis managers, first responders, and citizens alike as a series of 'what do we do now' problems triggered by the flow of events. These decision occasions emerge simultaneously or in succession over the course of the crisis [24, 25]. Protecting communities tends to require an interdependent series of crucial decisions to be taken in a timely fashion under very difficult conditions. Increasingly, there is a recognition that public sector resources (and traditional command, control capacities) are unlikely to suffice when dealing with the larger scale, more complex, and challenging contingencies. Recent experience from around the world clearly demonstrates the power of social media and personal communications-based information to empower and potentially improve decision-making and enable better access to expertise [26] and more agile, flexible, and decentralized forms of coordination and cooperation in crisis responses.

For example, during the terror attacks in Paris in November of 2015, citizens used the twitter hashtag #porteouverte (open doors) to seek and provide shelter to each other, so that individuals caught out in the open in affected neighborhoods could get off the street and out of danger [27]. Enabled by social media and smart devices, this potentially life-saving response was coordinated without the need for direct intervention from heavily burdened official responders.

Effective use of social media and smart devices are critical both for leveraging the potential for community-based response via self-organizing and for managing the interfaces between the public sector, private sector, and nonprofit sector components of a whole of community/society response (c.f. [28, 29]).

Meaning-making and Crisis Communication refers to the fact that crisis managers—across sectors—must provide relevant information in a timely fashion, attending not only to the operational challenges associated with a contingency, but also to the ways in which various stakeholders and constituencies perceive and understand it. Because of the emotional charge associated with disruptive events, followers look to leaders to help them to understand the meaning of what has happened and place it a broader perspective. By their words and deeds, leaders and other communicators can convey images of competence, control, stability, sincerity, decisiveness, and vision—or their opposites.

Social media channels—including direct social media-based communications by leaders on fora such as Twitter—have become a key arena in which information is exchanged and where alternative political visions as well as risk and situational assessments compete (c.f. [30–32]). A sound understanding of the discursive backdrop and the frames of reference of citizens and opinion leaders is essential to formulating and implementing effective strategies for crisis communication.

Before moving on, let us review some basics of leadership communication in the provocative context of contemporary crises: ([1]: 69–90; c.f. [33, 34]):

- *Credibility is a key asset; guard it!* Communicators who start out with or quickly develop credibility deficits face a significant additional obstacle with regard to crisis communication. By contrast, communicators who are proactive about getting and sharing the most salient information, promptly correct erroneous information, are circumspect about making and fulfilling promises will tend to maintain and even gain credibility over time. Credibility takes time to establish and rebuild, but can be destroyed in a single careless moment.
- *Crisis management is hard; manage expectations.* Crises are, by definition, difficult to manage. Distinctive features include value complexity and conflict, time pressure, and profound uncertainties regarding hazards and threats, efficacy or consequences of possible solutions, and reactions by adversaries, allies, other key stakeholders, and the public. Though it is often tempting to project optimism and impressions of control, recognizing the severity of the challenges to be faced and to be overcome tends to be a more prudent—and sustainable—posture. Acknowledging the seriousness of the situation and sketching out the steps being taken to prepare and respond tends to inspire—and is more likely to maintain—public confidence than a rosy scenario overtaken by events. Furthermore, lower citizen and media expectations are easier to fulfill.
- As noted above, *crises provoke strong emotions and stress* (for leaders and citizens alike). Crises are often associated not only with high levels of negative (e.g., fear, anger, outrage, shame, uncertainty) but also potentially with positive (courage, cooperation, pride, solidarity, focus) emotional states and expressions. Crisis communicators must seek to understand and adapt communication to the emotional states of those with whom they need to communicate.
- *Conveying crucial information is important and difficult—but not enough.* The first—and difficult enough—hurdle of crisis management is to make sure that everyone—inside and outside of government—has the information they need to play effectively their roles in the crisis management effort and/or to protect themselves and their loved ones. This sounds simple, but is in practice both vital and often difficult task. It is a necessary, but generally not sufficient, component of crisis communication.
- *Crises produce a demand for symbolic and emotional (as well as substantive) leadership.* Those experiencing negative emotions such as the ones mentioned in the previous point tend to look to their leaders (and others who communicate on their behalf) for hope, inspiration, empathy, and guidance. Citizens and employees expect leaders to inform themselves and recognize the importance of what has occurred, reach out to those who have suffered losses, affirm core community values, and point the way forward.

Accounting [1, 35] refers to the demands placed on crisis actors to justify their actions prior to, during, and in the aftermath of major crises and emergencies. The crisis literature questions are likely to be posed in a range of accountability fora such as:

- Why was it not possible to prevent the crisis from occurring or more effectively mitigate the damage?
- Why was the organization/society not better prepared?
- Why did delays, misunderstandings, mis-coordination, miscommunication, etc. occur?
- Why was the response not more effective, fair, legitimate, etc.?

The rise of the social media and personal communications technology in focus in the ATHENA project (and other related forms of public/personal surveillance technology such as closed-circuit television, police vehicle, body or helmet cameras, etc.) has profound impacts on and implications for the accountability process. The media and public sector accountability are now provided with real-time information, competing accounts regarding incidents, and feedback (though not necessarily sound, systematic, or reliable) regarding citizen reactions and satisfaction to services provided by crisis actors. In other words, the scope, complexity, granularity, and accessibility of accountability-relevant information have increased dramatically. This will be discussed in more detail in the following paragraphs.

Learning[2] from crisis requires an active, critical process which recreates, analyzes, and evaluates key processes, tactics, techniques, and procedures in order to enhance performance, safety, capability, etc. The learning process has just begun when the so-called lessons-learned document has been produced. In order to bring the learning process to fruition, change management/implementation must take place in a fashion that leaves the organization with improved prospects for future success [1, 21, 36–38].

The rise of social media, smart devices, and the internet has profound implications also for the task of learning from crisis. Traditional forms of journalism communicated through print or broadcast media have long been part of both practical and academic learning processes and environment as well as key empirical sources for crisis/post-crisis analysis. While both government sources and media sources have their gaps and biases, combining them has provided the analyst with more comprehensive accounts and means of identifying areas of controversy, contention, and competing narratives which may be subjected to source criticism.

The proliferation of social media and personal communications devices/technology (including photographic, video, and audio capability) means that crisis events and public, private, nonprofit, and citizen responses to them, are documented more quickly and more comprehensively than ever before. For example, the meteor event which struck the Russian town of Chelyabinsk on Feb 13, 2013, which is thought to have injured more than 1000 people, was documented in unprecedented detail by the dashboard cameras of local vehicles—providing extensive footage which could later be examined by scientists and other analysts to better understand the nature of the hazard [39].

In addition to individual and disaggregated citizen observations and media discourse, social media and GIS (Geographic Information Systems) enabled "Crisis

---

[2]The discussion of the implications of social media and smart devices for crisis learning draws heavily on Stern [36].

Mappers" provide maps documenting needs and resources and displaying them in spatial, scalable, and in formats enabling differential levels of detail ("zoomable"). These not only can contribute to enhanced situational awareness by government, NGOs, and citizens alike, but also provide alternative sources which can complement and be compared/contrasted with official accounts and crisis maps both during and after crises.[3]

Crises may be triggered or exacerbated by images or audio documenting apparent misbehavior or incompetence on the part of responsible authorities. For example, crowd control efforts on the part of the New York City Police Department during the Occupy Wall Street Protests were marred and public outrage stimulated by images of seeming police brutality against young female protesters. Similarly, social media posts documented the last moments of students, the seemingly counter-productive emergency management by the Captain and crew of the S. Korean ferry Sewol, and the relativity passivity of official responders on the scene before the ship sank [40]. Obviously, this phenomenon and the related spread of fixed video cameras in many public locations and places of business have important implications not only criminal prosecutions and deterrence of crime, but also for organizational/political accountability and learning processes (c.f. [1, 35]).

The often time-, place-, and date-stamped empirical material (though not immune to tampering, misinformation, and misrepresentation) produced by citizens via personal communications devices and published via social media provide additional material which can be used for developing and/or critically examining accounts depicting the unfolding of crisis events (c.f. [33]). Furthermore, the real-time nature and multi-directional communication provided by social media provide rapid feedback regarding community percepts of crisis management efforts potentially contributing to improved prospects for intra-crisis learning as well [37].

It should also be noted that various forms of open and closed community social media/web-based fora can provide infrastructure for crisis-related and ongoing social/organizational learning. As such, they can facilitate the development and operation of the so-called communities of practices in which experiences can be exchanged, problems and potential solutions discussed, and good practices shared. For example, the European Center for Disease Prevention and Control maintains an online and relatively open Field Epidemiology Manual Wiki which became an important resource for intra- and post-crisis learning related to the H1N1 Pandemic ([41]; https://wiki.ecdc.europa.eu/).

3.4 Summary and Prescriptive Reflections

Let us conclude this chapter by revisiting the key questions posed at the outset and formulating a few prescriptive reflections.

---

[3] http://crisismappers.net/

What is a crisis and how can a rigorous conceptualization of crisis help crisis managers (both official and citizen responders) to make sense of potential crises?

Though the concept of crisis can and should be defined in many ways for many different purposes, crisis can usefully be defined in terms of perceptions of :

- threats (and opportunities) with regard to core values cherished by citizens, communities, and, not least, their leaders.
- uncertainty with regard to threats and consequences of potential interventions
- urgency deriving from operational or political concerns as well as the demands of mass media

These criteria can be turned into a set of diagnostic questions that can help crisis responders from the public, private, and nonprofit sectors to orient themselves in the situation, target information gathering and analysis efforts, and help to make sense of an unfolding crisis [17].

What are some of the key and recurring leadership tasks associated with Crisis Management and how are they affected by the rise of social media and smart devices?

As we have seen, on the basis of extensive case research and review of the literature, it is possible to identify five core tasks of crisis leadership:

- Sense-making
- Decision-making and coordination
- Meaning-making (Crisis Communication)
- Accounting
- Learning

The review of these tasks found that the rise of social media and smart devices impact heavily on the conditions for and performance of all five core tasks.

1. *The Social media/smartphones combination can work for or against crisis managers:* There can be no doubt that social media is a potentially powerful force multiplier and has rapidly become an essential tool in the contemporary crisis management tool box. However, it is important to be aware that social media and the capabilities of smart devices can work for—or against—any given organization at any given time. These capabilities can be used not only by 'pro-social' actors but also in various ways by foreign state adversaries, terrorists, criminals, and others who do not have society's best interests at heart or exhibit poor judgment in the face of threat. Organizations need to be prepared to use social media proactively and offensively—to seek to gain and maintain the initiative and a high level of situational awareness—in crisis situations. They also need to be prepared to cope with social media-based propaganda and disinformation. In today's communications environment crisis managers must be alert and resilient enough to ride out viral waves—and seek to turn the tide—of misinformation, outrage, and other negative forms of social media reactions.
2. *Mind the digital divide and embed social media strategies in a comprehensive approach to risk communication and crisis management.* Though social media use

continues to grow and spread across demographics in society, it is important to keep in mind that that there are and are likely to continue to be significant elements of the population who choose to refrain from—or lack the means and/or know how—to make use of social media and the smartphone revolution. This has a number of aspects. First of all, communication strategies should make use of a variety of modalities and differentiated approaches to reach the full range of target groups in society. Analog (e.g., warning sirens and loudspeakers) as well as digital means, use of conventional as well as social media, may all have their place in crisis communication strategies. Furthermore, crisis communication is facilitated by having effective risk communication and issues management programs well before (and after) periods of acute crisis [42, 43]. Cultivating credibility and educating the public and the media in advance about complex risks, set the stage for effective crisis communication and a fruitful conversation under crisis conditions.

3. *Be open and forward-looking with regard to emerging applications, platforms, and devices.* While it may be tempting to tailor social media tools and strategies to whatever forms of social media and devices are currently the most popular, it is important to take a more open and flexible approach. New forms of general and specialized social media emerge (and fade) regularly. Social media components of crisis management strategies should be designed to be able to adapt to and make use of not only current devices and social media platforms but also to incorporate others likely to emerge in the future. While initially it may make sense to emphasize the biggest 'players' (e.g., platforms like Facebook and twitter) mature social media strategies will match media formats and capabilities with more specific purposes and target groups.

4. *Embrace the multi-directional and multi-dimensional character of social media and smart devices in developing short- and long-term crisis management capacity.* While highly effective for that purpose as well, social media and smart devices are not merely useful for pushing out information. Social media can be used to gather information, enable analysts and decision-makers and connect them to communities of practice (c.f. [26]), assess reactions to organizational messaging, and provide an enhanced basis for post-crisis accountability, learning, and reform.

Acknowledgements The research reported in this chapter was supported by the EU (7th Framework Program Project ATHENA), the Swedish Civil Contingencies Agency, and the Swedish Defense University. The author would like to thank a number of colleagues for intellectual and editorial input to various sections including Babak Akghar, Andrew Staniforth, Dave Waddington, Ellie Lockely, Eva Karin Olson, and Stephanie Young.

References

1. Boin, A., t'Hart, P., Stern, E., & Sundelius, B. (2005; 2016 in press). *The politics of crisis management: Public leadership under pressure.* New York, NY: Cambridge University Press.
2. Helsloot, I., Boin, A., & Jacobs, B. (Eds.). (2012). *Mega-crises.* Springfield: Charles Thomas.

3. OECD. (2015). *Adapting government approaches to a new crisis landscape. The changing face of strategic crisis management. OECD reviews of risk management policies*. Paris: OECD Publishing.
4. Rosenthal, U., Boin, A., & Comfort, L. (Eds.). (2001). *Managing crises*. Springfield: Charles Thomas.
5. Akghar, B., & Yates, S. (Eds.). (2013). *Strategic intelligence management*. Oxford: Butterworth-Heinemann.
6. ATHENA 3.1 Report. (2014). Human factors: Mapping constraints on sense-making and communication in crisis situations.
7. Rosenthal, U., Hart, P., & Charles, M. (1989). From case studies to theory and recommendations: A concluding analysis. In U. Rosenthal, M. Charles, & P. t'Hart (Eds.), *Coping with crises: The management of disasters, riots, and terrorism* (pp. 436–472). Springfield, IL: Charles C. Thomas.
8. Stern, E. (2003). Crisis studies and foreign policy analysis. *International Studies Review, 5*, 183–202.
9. Hermann, C. F. (1963). Some consequences of crisis which limit the viability of organizations. *Administrative Science Quarterly, 8*, 61–82.
10. Farnham, B. (1997). *FDR and the Munich crisis*. Princeton: Princeton University Press.
11. Hermann, M. G. (1979). Indicators of stress in policymakers during foreign policy crises. *Political Psychology, 1*(1), 27–46.
12. Holsti, O. R. (1989). Crisis decision making. In P. E. Tetlock et al. (Eds.), *Behavior, society, and nuclear war* (pp. 8–84). New York: Oxford University Press.
13. Flin, R. (1996). *Sitting in the hot seat: Leaders and teams for critical incident management*. New York: Wiley.
14. Post, J. (1991). The impact of crisis-induced stress on policy makers. In A. L. George (Ed.), *Avoiding war*. Boulder: Westview Press.
15. Weick, K. E. (1995). *Sense making in organizations*. Thousand Oaks, CA: Sage Publications.
16. Klein, G. (2001). *Sources of power: How people make decisions* (7th ed.). London: The MIT Press.
17. Stern, E. (2009). Crisis navigation. *Governance, 22*(2), 189–202.
18. Keeney, R. (1992). *Value-focused thinking*. Cambridge: Harvard University Press.
19. Mitroff, I., & Silvers, A. (2009). *Dirty rotten strategies: How we trick ourselves in solving the wrong problems precisely*. Palo Alto: Stanford Business Press.
20. George, A. L. (1980). *Presidential decision making in foreign policy: The effective use of information and advice*. Boulder, CO: Westview Press.
21. Stern, E. (2015). *Understanding and identifying strategic crises through early warning and sensemaking. The changing face of strategic crisis management. OECD reviews of risk management policies*. Paris: OECD Publishing.
22. Weick, K. E. (1988). Enacted sense-making in crisis situations. *Journal of Management Studies, 25*(4), 305–317.
23. Alberts, D. S., & Hayes, R. E. (2003). *Power to the edge*. Washington, DC: Command and Control Research Program.
24. Stern, E. (1999). *Crisis decisionmaking: A cognitive-institutional approach*. Stockholm: Department of Political Science, Stockholm University.
25. Stern, E., Deverell, E., Fors, F., & Newlove-Eriksson, L. (2014). Post-mortem crisis analysis: Dissecting the London bombings of July 2005. *Journal of Organizational Effectiveness., 1*(4), 402–422.
26. Koraeus, M., & Stern, E. (2013). Exploring the crisis management/knowledge management nexus. In B. Akghar & S. Yates (Eds.), *Strategic intelligence management* (pp. 134–149). Oxford: Butterworth-Heinemann/Elsevier.
27. McHugh, H. (2015). After Paris attacks Parisians use hashtag to offer shelter. *Wired 11/13/15*. Retrieved March 24, 2016, from http://www.wired.com/2015/11/paris-attacks-parisians-use-porteouverte-hashtag-to-seek-offer-safe-shelter/

28. Hughes, A. L., & Tapia, A. (2015). Social media in crisis: when professional responders meet digital volunteers. *Journal of Homeland Security and Emergency Management, 12*(3), 679–706. doi:10.1515/jhsem-2014-0080.
29. Kaufman, D., Bach, R., & Riquelme, J. (2015). Engaging the whole community in the United States. In R. Bach (Ed.), *Strategies for supporting community resilience*. CRISMART (Vol. 41, pp.151–186). Stockholm: Swedish Defense University.
30. Sutton, J. (2009). Social media monitoring and the democratic national convention: New tasks and emergent processes. *Journal of Homeland Security and Emergency Management, 6*(1), 1–20.
31. Sutton, J., Palen, L., & Shklovski, I. (2008) Back-channels on the front lines: Emerging use of social media in the 2007 Southern California Wildfires. In *Proceedings of the 5th International ISCRAM Conference*, Washington, DC.
32. Wukich, C. (2015). Social media use in emergency management. *Journal of Emergency Management, 13*(4), 281–294.
33. Olsson, E. K. (2014). Crisis communication in public organizations: Dimensions of crisis communication revisited. *Journal of Contingencies & Crisis Management, 22*(2), 113–125. doi:10.1111/1468-5973.12047.
34. Veil, S., Buehner, T., & Palenchar, M. (2011). A work in process literature review: Incorporating social media in risk and crisis communication. *Journal of Contingencies & Crisis Management, 19*(2), 110–122.
35. Boin, A., McConnell, A., & t'Hart, P. (Eds.). (2008). *Governing after crisis*. New York, NY: Cambridge University Press.
36. Stern, E. (2015). Bridging the crisis learning gap. From theory to practice. In N. Schiffino, L. Taskin, C. Donis, & J. Raone (Eds.), *Organizing after crisis: The challenge of learning* (Chapter 11). Berlin: Peter Lang Publishing Group.
37. Deverell, E., & Olsson, E.K. (2009) Learning from crisis: A framework of management, learning and implementation in response to crisis. *Journal of Homeland Security and Emergency Management, 6*(1), Article 85.
38. Stern, E. (1997). Crisis and learning: A conceptual balance sheet. *Journal of Contingencies & Crisis Management, 5*(2), 69–86.
39. The Guardian. (2013, November 6). *Scientists reveal the full power of the Chelyabinsk meteor explosion*. Retrieved June 1, 2014, from http://www.theguardian.com/science/2013/nov/06/chelyabinsk-meteor-russia
40. New York Times. (2014, May 1). Retrieved June 1, 2014, from http://www.nytimes.com/2014/05/01/world/asia/korean-ferry-students-captured-sinking-on-video.html?action=click&contentCollection=Asia%20Pacific&module=RelatedCoverage&region=Marginalia&pgtype=article
41. Greco, D., Stern, E., & Marx, G. (2011). *Review of ECDC's response to the influenza pandemic 2009/10*. Stockholm: European Center for Disease Prevention and Control.
42. Drennan, L., McConnell, A., & Stark, A. (2015). *Risk and crisis management in the public sector*. London: Routledge.
43. Regester, M., & Larkin, J. (1997). *Risk issues and crisis management*. London: Kogan Page.

Chapter 4
Case Studies in Crisis Communication: Some Pointers to Best Practice

Kerry McSeveny and David Waddington

4.1 Introduction

Increasing use of social media has significantly altered the nature of the role played by ordinary citizens in crisis situations, and has consequently changed the dynamics of the relationship between the general public and those involved in leading the response effort (e.g. [1]). Social media can play a central role at all stages of a crisis situation, from planning and preparedness before an event, providing situational awareness and coordinating response during the event, and facilitating recovery in the aftermath [2]. Crisis involves 'a serious threat to the basic structures and the fundamental values and norms of a system which under time pressure and highly uncertain circumstances necessitates making vital decisions' ([3], p. 5).

During such an event, social media platforms can play an essential role in facilitating this decision-making, through crowdsourcing information, and enabling vital communication between crisis managers and the public. However, research has shown that there are a number of barriers to the widespread adoption of social media amongst official responders, including lack of resources and other institutional barriers, challenges with verifying information generated via these channels, and lack of confidence in using social media [4, 5], so it is essential to consider how this potential might best be harnessed in the management of crisis incidents (See Chaps. 3, 5, and 6).

While there are existing guidelines for emergency managers regarding the use of social media in emergencies (e.g. [6]), this chapter takes a case study approach, using a more detailed analysis of specific events to more closely examine the risks and benefits of using social media in crisis communication, and to identify examples of best practice in this regard. We consider the use of social media by first responders and government officials in a range of crisis situations which took place

K. McSeveny (✉) • D. Waddington
CENTRIC, Sheffield Hallam University, Sheffield, UK
e-mail: K.McSeveny@shu.ac.uk; D.P.Waddington@shu.ac.uk

© Springer International Publishing AG 2017
B. Akhgar et al. (eds.), *Application of Social Media in Crisis Management*,
Transactions on Computational Science and Computational Intelligence,
DOI 10.1007/978-3-319-52419-1_4

in 2011, with a focus on public disorder (riots in Manchester and London, UK), terrorist activity (the attacks in Oslo and Utoya, Norway), and natural disasters (the earthquake and subsequent nuclear incident in Tohoku, Japan).

4.2 The Case Studies

Each case study draws together existing research and analysis relating to the use of social media during the event in question, and reflects on the lessons learned, with a particular focus on communication between officials and the general public (See Chap. 13 for Athena project final test case). The number of possible case study examples in these areas is vast, but these particular events have been selected due to their contemporary significance, the quantity and quality of academic analysis currently available, and their subsequent potential to provide valuable insight into the potential role that may be played by social media in crisis preparedness and communication. Analysis of the case study examples uses a combined method informed by the approaches of 'process tracing methodology' and 'structured focus comparison' outlined by Stern and Sundelius [7]. This approach involves the systematic reconstruction of the events surrounding a crisis (often from a wide variety of sources), and comprises four main steps: (1) Placing the crisis in its historical, institutional, and political context; (2) Reconstructing the precise course of events; (3) Identifying key points of decision-making; and (4) Conducting cross comparison between cases to enable the identification of best practice.

Using accounts available from a range of sources, including research articles, media reports, and policy documents, each case study in turn sets out the contextual background and main sequence of events that constituted the crisis in question, establishing the ways in which social media was utilized by various parties in relation to key decision-making occasions. The case studies are then subjected to an overall thematic analysis which is presented as a series of recommendations for the use of social media in crisis communication, with regard to communication between crisis managers (including emergency services and other authorities) and the citizens affected by an incident.

4.3 Social Media and Public Disorder: The English Riots, 2011

4.3.1 Context

An examination of the use of social media and mobile devices during the English riots of August 2011 reveals the key role played by social media, both in the communication between members of the public, and in the attempts of police forces to prevent or contain disorder. There has been no shortage of examples of the ways in which various forms of social media and mobile devices have been employed to help mobilise huge gatherings of people and/or entire social movements (see, for example, [8, 9]). However, the selection of the English riots for the purpose of case study is a reflection, not only of their relative recency and contemporary significance, but also of the fact that they represent virtually the first major example of rioting in which the use of social media has not only been prevalent but also subjected to close academic analysis.

The catalyst for the series of riots which took place in August 2011 in cities across England centred around the killing of Mark Duggan, a 29-year-old African Carribean man, who was shot dead on 4 August, close to his home in Tottenham, North London, by members of the Metropolitan Police Service's Operation Trident. The anti-gun crime unit had been following him with the intention of arresting him on suspicion of carrying a firearm. Although the precise sequence and nature of events remain highly contested, the rumour quickly spread that Duggan had been callously 'assassinated', while offering scant resistance, by one or more members of the Trident team [10]. Questions about whether or not Duggan shot first and whether this was an act of self-defence started a debate that put the police operation into question. On the evening of 6 August a crowd of about 300 people gathered at a police station. What started as a peaceful demonstration turned into a forceful riot that spread in the following days across neighbourhoods and to other cities such as Birmingham, Liverpool, and Manchester. Buildings were set on fire and stores were looted, and thousands of people were arrested. Five people died and over 200 people were injured; 186 of them police officers. In London alone, 3443 riot-related crimes were reported which caused damages of over 200 million pounds.

4.3.2   Public Use of Social Media

During the riots, social media became a contentious topic of public debate. While addressing an emergency session of the UK Parliament, convened in the aftermath of the four consecutive nights of rioting and looting, the British Prime Minister, David Cameron, highlighted the role supposedly played by social media (especially Twitter) in the instigation and development of the disorders [11]. This view was exemplified in the subsequent prosecution of a number of individuals, including two young men who were charged with 'online incitement' and each jailed for 4 years, for the creation of Facebook events and pages encouraging rioting behaviour, despite the fact that no riots ensued as a result [12]. The Prime Minister was quoted as having approved of the 'tough stance' adopted by the courts.

However, although Twitter and Facebook were widely cited as playing a pivotal role in the rioting, analysis of patterns of communication during these events reveals that it was in fact BlackBerry Messenger (BBM), a system which allows publicly untraceable dissemination of 'one-to-many' messages by individuals to a network of contacts, which was utilised in organising and inciting the unrest [13]. This is supported by an analysis of the Guardian newspaper's database of over 2.5 million riot-related tweets which reveals that although disorder initially broke out in

Tottenham between 8 p.m. and 9 p.m. on Saturday night, Twitter activity was practically non-existent in the hour preceding the disorder, with only 15 tweets recorded
between 7 p.m. and 8 p.m., compared with a total of 15,000 between 10 p.m. and 11
p.m. [11]. Regardless of the specific nature of the communication methods used,
Baker, amongst many others, advises extreme caution when attributing blame for
the unrest to the technology itself:

> To blame technology as the cause of the riots is accordingly limited. Riots have occurred at
> regular intervals in modern Britain long before these technological innovations, and while
> new social media facilitates social networking in diverse temporal and spatial boundaries,
> it is a facilitator rather than the underlying cause of collective action ([14], p. 45).

As well as challenging the misconception that Twitter was implicated in mobilising the rioters, the 'Reading the Riots' study (collaboratively undertaken by journalists on the Guardian newspaper and academics from the London School of
Economics) provides evidence that social media played a much more prosocial role
in the events of August 2011. An analysis of 2.6 million tweets reveals that Twitter
was mostly used for sharing news, moderating the spread of rumours, and facilitating the organisation of volunteers to deal with the aftermath of the rioting [15, 16].
These 'cleanup' operations were widespread, mobilising more than 12,000 people
around the hashtags #riotcleanup and #riotwombles [17, 18].

Social media was also used in the efforts to identify perpetrators [19], and the
general reaction of Twitter users to messages allegedly inciting them to riot was
overwhelmingly negative, with many recipients to forwarding such tweets and
details of their sources on to the police [18]. However, despite this extremely positive picture, it is important to note that not all of these contra-riot activities were
unequivocally prosocial in nature. For example, Tonkin et al. [20] report on the
emergence of 'vigilante groups', including a group of Millwall Football Club sup-
porters who took to the streets in opposition to the rioters.

Twitter also operated to moderate social alarm and suppress the possibility of
further public disorder by debunking circulated rumours, including notable fabrica-
tions relating to a fire on the London Eye, the release of a tiger from London Zoo,
and an attack on Birmingham Children's Hospital. Despite this self-regulation,
there is still a need for the timely dissemination of accurate information and advice
from trusted social media sources, and social media users' reliance on reliable
sources can be observed in the tendency to retweet information that came from
mainstream news outlets (such as the BBC, ITV, and Channel 4 news hashtags),
prominent celebrities or media figures, and (though slightly less often) the police
themselves [20, 21].

4.3.3   Police Use of Social Media

The Reading the Riots researchers observed that, in general terms, the police failed
during the riots to take full advantage of the social media intelligence and networking capacity potentially available, and they saw the police use of social media as

having been hampered by inadequate resourcing and the lack of a coherent strategy [16]. It is possible that in at least one important respect, concerning their failure to quash the growing local rumour that Mark Duggan had been 'assassinated' by the pursuing officers, the police were culpable of not employing social media when it would have been highly advisable to do so [22].

At the time of the rioting, police use of social media in the UK was variable between forces, and this inconsistency remains the case to date [23]. An evaluation of the 'Engage' police community engagement strategy revealed that 30 police forces had established a corporate Twitter account, alongside a total of 140 neighbourhood and local policing team accounts, with some forces adopting social media more enthusiastically than others [24]. Analysis of the way that forces used social media noted the tendency for the accounts to be used for the one-way dissemination of information from the police to the public, with fewer examples of two-way engagement (e.g. [24, 25]).

This variability in individual forces' use of social media was also evident during the riots, exemplified in the contrast between the activities of the police forces in two of the cities which experienced unrest during this period—the London Metropolitan Police (the Met) and Greater Manchester Police (GMP) [15, 16]. One obvious advantage in the GMP's favour was that they were already able to depend on the existence of a well-established social media infrastructure, involving 60 localised accounts, all of which were extremely active down at community level. In comparison, a senior Met officer acknowledged in the wake of the riots that: 'We're still not wholly up to speed in using social media as an intelligence tool, an investigative tool and most importantly as an engagement tool' (ibid., p. 21). Twitter usage statistics also reveal the extent of this disparity—during the period from 4 to 13 August, the Met posted 132 tweets, but GMP almost three times as many (a total of 371). While the Met increased their number of Twitter followers from 4000 to 42,000 with the onset of the riots, GMP increased their following from 23,000 to 100,000, making it second only to the FBI as the world's most popular police force on Twitter [26].

There was a further major difference in terms of the content and style of messages posted. Both forces employed Twitter with a primary objective of gathering and disseminating information about the riots (e.g. to seek information on offenders by posting CCTV images of perpetrators on Flickr and leaving phone numbers and website addresses by which to provide relevant details), but the GMP placed a much greater emphasis on reassuring the public and addressing rumours. In contrast, the Met clearly favoured a much more impersonal style, directed to a generic audience, with a primary focus on maintaining law and order:

> MET: The Met has now arrested 1401 people in connection with violence, disorder and looting. 808 of these have been charged (Denef et al., [26] p. 3474)

The GMP took a more personalised approach, taking the time to reply to individual messages, addressing specific comments and queries:

> USR1: @gmpolice is it true that chaos has started in town, carphone warehouse has been done over already??

GMP: @USR1 nothing at the moment follow us and we will let you know if there is anything to report (ibid., p. 3476)

This more engaged approach also extended to the GMP responding to criticism for the tone of some of their messages, apologising for and retracting the contentious messages, and explicitly asking the public for feedback on how they were handling the situation on Twitter. Comparisons of public response to the social media approaches applied by the two forces over this period highlighted the relative success of the GMP's more 'expressive' approach compared to the Met's 'instrumental' strategy [15, 16, 26].

4.4 Social Media and Terrorist Incidents: The Oslo and Utoya Attacks, 2011

4.4.1 Context

Social media provides a potentially valuable resource for officials and emergency responders as a means of gathering intelligence on extremist groups, identifying possible terrorist incidents before they occur, and for coping with incidents as they happen [27]. However, research focusing on the management of specific instances of terrorist activity suggests that social media is not always utilised as effectively as it could be for communicating with the public under these circumstances (e.g. [28, 29]). The evidence suggests that clear social media strategies are not yet widely implemented by emergency managers and/or the authorities during such attacks, and the 'void' left by this lack of official social media engagement can have significant consequences. This is particularly evident in the second case study in this chapter, which focuses on the atrocities carried out in the Norwegian capital of Oslo and the island of Utoya by right-wing extremist Anders Behring Breivik.

On 22 July 2011, Breivik detonated a car bomb near the Prime Minister's office and other government buildings, an act which resulted in eight fatalities and serious injuries to 30 other people. Breivik then moved on to the island of Utoya, where (disguised as a police officer) he proceeded to open fire on scores of young people attending an annual worker's youth league summer camp. Sixty-nine more lives were lost as a consequence of these shootings and a further 60 individuals were wounded. A little over a year later, Breivik was found guilty of the mass killings associated with both attacks. The jury had heard how a 1500-word statement appearing on Breivik's Facebook page, just before the attacks occurred, had alluded to a vaguely articulated but nonetheless possible motive for his actions [30, [31].

It is evident from subsequent accounts of this atrocity that the response of the police and other relevant authorities was both haphazard and indecisive—especially in relation to the Utoya shootings, where police attempts to reach the island were strongly criticised for having been too slow, and widely ridiculed because they initially set out aboard a boat with a defective engine ([31], op. cit.). Above all, however, it was the poor use of crisis communication by the police which attracted the greatest amount of criticism, as a combination of short staffing and computer malfunction meant that the Norwegian government's crisis communication secretariat was not able to respond as effectively or promptly as was necessary. It is the primary lessons to be drawn, not only from these shortcomings of police communication, but also in the way in which social media interaction between citizens immediately began to fill this communicative vacuum, that we address in our discussion.

4.4.2   Public Social Media Use

The absence of practical information and reassurance from the police and other authorities led to the emergence of processes of 'resilience governance', where citizens helped one another to make sense of events and share information. Perng et al. (op. cit.) make the justifiable point that it was due to this scarcity of insightful, reassuring and practically useful communication from the police and other emergency response agencies that the widespread citizen use of social media almost instantly emerged as if to compensate for the absence of official information. A complementary study by Kaufmann [32], based on in-depth interviews with 20 people who used social media during and after the Norwegian atrocities, emphasises how there was a tremendous need on all their parts to manage a shifting spectrum of high emotions, from 'shock, confusion, disbelief, worry, and fear at first … followed by the urge to access information, which was accompanied by a sense of sensationalism' (ibid., p. 10).

Kaufmann (ibid.) documents the ways by which existing 'social media infrastructures' and users worked together in such a way as to initiate and mediate various processes of 'resilience governance' which helped people both to understand and deal with the shocking and tragic events. These activities included the use of social media to determine what exactly was happening, through the exchange of information and knowledge, and to make sense of the situation through the organisation of Facebook events to 'meet up' and discuss events with a view to obtaining clarity and reassurance. Social media was also widely used to spread reassuring messages regarding individuals' safety and well-being, a process which was formalised through the creation of a Facebook event, 'Oslo er i god behold', on which over 64,000 subscribers attested to the fact that they were free of harm. Social media also provided a space for 'therapeutic sharing and digital mourning' (ibid., p. 11), and presented opportunities for political and moral standpoints to be expressed and their merits discussed. Kaufmann notes the significant role that social media played in engendering a mediated form of 'community and unity', but cautions that respondents found this online interaction problematic in some regards, citing information overload, concerns about authenticity, and discomfort about the sensationalist and overly personal nature of some of the exchanges.

A more immediate and potentially more catastrophic consequence of the public's use of social media in crisis situations is raised by Perng et al. (op. cit.). They observe how the lives of some of those people who were trapped and in danger on the island were placed in even more jeopardy by the transmission of socially mediated messages. For example, shortly after the shooting had started on Utoya, one social media user retweeted a message from one of the people on the island:

> 'We are sitting by the lake. A man dressed in police uniform is shooting. Help us'.

However, it soon became apparent that this message, and subsequent updates which also drew on first person reports from the island, might well encourage concerned (and even desperate) friends and relatives to try calling such victims on their mobiles, whose ringing could expose their locations to the gunman and put them in danger. Consequently, @*NilsPetter* was obliged to hurriedly urge his followers:

> 'DO NOT CALL acquaintances on Utoya! It can put them in danger. Wait until they call you, even if it is bloody unbearable!' (ibid.)

Similar unforeseen risks surrounded the subsequent attempts by concerned onlookers to use Twitter messages as a way of coordinating rescue attempts by boats. Though undoubtedly invaluable and well intentioned in principle, the practical effect of such initiatives was often to endanger the would-be rescuers and serve to confuse and complicate the efforts of the police and other emergency services.

4.4.3   Official Communications and Use of Social Media

Evaluation of the crisis management during the events of 22 July highlights a series of poor decisions and coordination failures, including ineffective mechanisms for the dissemination of information from senior level, combined with a lack of local communications expertise in the Utøya area and the district's police force. The lack of clarity around the situation was exacerbated by the need for any available information to go through multiple levels of authorisation before reaching those responsible for making decisions, meaning that 'information from the police to the Council, the MJ [Ministry of Justice] and the Cabinet, and eventually to the public, was slow and inadequate. Media reports were more up to date than the information emanating from government' ([33], p.361). Likewise, Falkheimer [30] notes that the speed at which information can be disseminated via social media and modern mass media necessitates a much more flexible approach to crisis communication which enables improvisation.

   In an analysis of the specific ways in which the official communication strategy proved inadequate for the situation, Falkheimer (op. cit., p. 58) makes the point that the two attacks in question were carried out at a time when the Norwegian government's national crisis communication secretariat was temporarily under-strength, as due to staff vacation only three of its eight regular staff members were on duty. Compounding this problem was the fact that the need to change location, combined with technical malfunctions preventing access to both of the internet and of their own intranet facility, meant that they were unable to access established online emergency plans and strategies. Correspondingly, the Police Department's crisis communication plans were criticised for being 'outdated and not integrated in the organizational work' (ibid. p. 58).

Despite these limitations, crisis communication staff felt that they managed to overcome some of these problems, using an email loop between the key offices and organisations, including the secretariat, the police, the Ministry of Justice, and other involved parties. This email loop was used to coordinate regularly updated 'speaking points' which were distributed to the Prime Minister's temporary office, and were used as the basis for the preparation of the Prime Minister's speech. Falkheimer notes that 'The main problem, according to the Norwegian crisis coordinators, was the lack of information flow from emergency organizations, especially the police, and the national government' (ibid., p. 58).

A subsequent official evaluative report of police operations during the crisis was extremely critical of the crisis communications involved (Falkehimer, ibid.). The evaluators were particularly condemning of the police's failure to correct initially false rumours that only ten people had been killed in the shootings (when the actual figure was far higher), and that the alleged perpetrators of the atrocities were members of Al Qaeda. Equally strong criticism focused on the lack of coordination between the central police department in Oslo and the police district responsible for the disaster area of Utoya, and on the fact that there was a virtual absence of any communication infrastructure within this crucial locality. In summary, 'the need for personnel was not met (in order to meet the communication needs), the plans were poorly updated, failure to ensure that all districts had communication staff, failure to practice crisis handling more, failure to sufficiently communicate via social media, and that they were too late in the disclosure of information in general' (ibid., p. 59).

These communication issues, and in particular the failure to make use of social media, were widespread. In fact, the only notable instance of the use of social media by the emergency services involved an attempt by the Oslo University Hospital/ Ullevål to issue an urgent humanitarian message via Twitter, in an attempt to deal with a predicted shortage of a specific blood type as a result of dealing with an unusually high number of casualties. However, their request 'Oslo University Hospital needs blood donors with blood type O—(O negative). Ring the blood bank now—telephone 2211 8900 or 2211 8865' failed to specify that only registered blood donors could donate (quoted in [35], p. 97). Although this omission was corrected in a follow-up message by administrators, this was too late to prevent the unnecessary arrival (and inconvenience) of many unregistered donors. However, it does constitute an all-too-rare example of the effective use that can be made of social media communication by the emergency authorities.

It is important to note that despite the serious shortcomings of the crisis communication during the atrocities themselves, the Norwegian authorities, including government officials and royal family, have been praised for their handling of the immediate aftermath of the terror attacks. Their responses were widely perceived as those of 'fellow citizens', inspiring confidence while showing care and compassion towards the Norwegian people. Prime Minister Jens Stoltenberg's attitude in particular was interpreted as 'inclusive, fatherly and emphatic' ([34], p. 59).

Examination of the Norwegian authorities' use of social media during the more recent 2014 terror alert suggests that there are still improvements to be made in the event of a crisis situation. Rasmussen and Ihlen [36] observed that throughout the duration of the alert official social media messages tended to take the form of one-way communications, often offering 'vague and abstract' information, or containing 'professional jargon'. Their discussion also highlights some of the fundamental operational difficulties in administrating official social media accounts, namely maintaining a consistent 'voice' for such accounts, noting that the National Police Directorate supports 27 police districts, each with their own Twitter account, and the Oslo Police District's Twitter account is maintained by twelve individual operators. They suggest caution against the adoption of too informal a tone in professional police communications, suggesting that likability or popularity should not be sought at the expense of police authority.

4.5 Social Media and Natural Disasters: The Tohoku Earthquake

4.5.1 Context

At 14:46 JST on 11 March 2011, an earthquake measuring a magnitude of 9.0 struck the northeast area of Japan, with its epicentre in the Pacific ocean, approximately 70 km from the Oshika Peninsula of Tohoku. The Tohoku earthquake (also known as the Great East Japan earthquake) was the most powerful recorded in Japanese history, and triggered a tsunami which engulfed the northern part of the country. The tsunami was detected by the early warning system, but at 40 m above sea level it was much higher than experts expected, so damage was widespread, affecting over 30 cities ([37], p. 14).

Speaking about the disaster, Prime Minister Naoto Kan said that 'The earthquake, tsunami and the nuclear incident have been the biggest crisis Japan has encountered in the 65 years since the end of the second world war' (quoted in [38]). The earthquake and subsequent tsunami were devastating, resulting in over 15,000 deaths, 6000 injuries, and 2500 missing people. Over 220,000 people were forced to leave their homes, and there was widespread damage to buildings, power and water supplies, communications, and transport infrastructure. The tsunami also hit Fukushima Daiichi nuclear power station, causing a cooling system failure which led to explosions in three of the nuclear reactors and necessitated evacuation of an area with a radius of 20 km from the plant [40]

The Japanese government was quick to respond, and drew on existing disaster prevention and emergency management systems, informed by experience of previous natural disasters in the country. This approach 'encouraged local governments to strengthen their responsibility as first responders and established e-learning systems to educate employees on the use of communication technologies and networks for disaster prevention and emergency management' ([39], p. 32).

However, despite the overall effectiveness of the government's response and relief efforts, they were strongly criticised for their failure to provide accurate information to their citizens, particularly in relation to the nuclear incident [40]. In this third and final case study, we will therefore explore some of the issues surrounding the government's use of social media, in order to learn more about why their communication strategies may have been considered ineffective in this regard.

4.5.2   Public Use of Social Media

A number of recent emergency situations have demonstrated the vital role that social media can play in as a communication channel in the event of a natural disaster, and studies of social media use during crises such as floods, cyclones, and wildfires note the focal role that 'official' social media accounts play during natural disasters, observing significant increases in follower numbers for Twitter and Facebook accounts associated with local police forces, fire services, and other emergency responders, as well as huge numbers of message likes, retweets, and other forms of engagement (e.g. [41, 42]). Reports of the occurrences of March 2011 reveal that online channels were key to 'initiating search/rescue operations, fundraising, providing emotional support, and creating, delivering and sharing information' as events unfolded ([39], p. 37).

During the earthquake a vast range of online platforms were used for different purposes and to achieve distinct goals, locally (in the affected Tohoku area), nationally (across Japan), and globally. This case study will focus mainly on the Japanese government's use of social media, chiefly Twitter, to communicate with citizens affected by the earthquake and subsequent events, and to disseminate official messages amongst the population, but a wide range of online social media was used to fulfil a diverse number of roles and functions [43]. A possible reason for this is the robust nature of social media—online channels, particularly when accessed via mobile devices, may also withstand conditions that other communication methods, such as telephone services, do not. Research shows that over half of directly affected individuals, and around 80% of indirectly affected individuals, cited social media as the source they most relied on for information during the earthquake ([44], p. 15).

One significant feature of social media is its immediacy—sites such as Twitter and Facebook can be used to quickly disseminate news. In this case, news of the event was broken on Twitter around 20 min before it was reported by mainstream media. As some messages were posted in English, news of the earthquake spread rapidly beyond Japan and across the world ([39], p. 29). Twitter quickly became a means to exchange information, and to disseminate disaster-related news, and the number of tweets-per-minute in Japan increased from 3000 to 11,000 during the earthquake ([45], p. 29).

Online platforms, including Twitter, Mixi, and Google Person Finder, were used by survivors to contact friends and family to provide reassurance that they were safe. Likewise, content sharing sites like Youtube and Ustream were used to disseminate information and share experiences. The social networks also took direct action to assist the relief effort, providing links to relevant websites, and giving information in both Japanese and English [44, 45]. Although Facebook was not yet very popular in Japan, its position as the largest social network in the world meant that it played a vital role in communicating information about the disaster worldwide. This enabled messages of support and relief donations to be sent from around the globe, mobilised in part by the social network's group functionality, and a dedicated page called 'Disaster Relief' received over 680,000 followers and 'likes' [44]. The Prime Minister's Office set up a Twitter account (@Kantei_Saigai) and a Facebook account during the days following the earthquake, responding to international interest in the events by publishing English translations of official press updates and briefings ([45], p. 31).

A number of Wikis and ad hoc information sharing services were set up by volunteers, who used Google, Yahoo, and Mixi to compile and share data, providing lists of shelters, locating missing people, and sharing information about transport infrastructure, amongst countless other functions. These services were mostly unofficial and citizen led, and complemented official relief work ([46], p. 2). Sinsai.info, a 'crisis mapping' tool built during the disaster, used crowdsourced information alongside online maps (such as Google Maps and Google Earth), and is based on the Ushahidi platform developed in response to the 2010 Haiti earthquake [44].

While social media played a valuable role in communication and information sharing during the earthquake, there are limitations to its potential in these contexts, as many in the areas most affected by the earthquake did not have mobile or internet access for several weeks [45]. Those who could use social media channels reported issues relating to the reliability of information obtained online. The persistence of false rumours, combined with the sheer volume of messages circulating, made identifying genuine warnings or calls for help extremely difficult, and resulted in unnecessary stockpiling of essential supplies [47]. Other, more malicious tweets were designed to extort money under the guise of the collection of relief donations [48].

While data mining of Twitter feeds has been identified as a possible means to more efficiently target relief efforts [48], this was always not borne out by experience. Although it provided an extremely useful platform for the public to locate missing people and obtain up-to-date information, humanitarian organisations and local officials found the information provided on Twitter and Facebook less helpful for identifying where assistance was needed. The Japanese government's efficiency in providing aid, combined with the difficulties arising from messages circulating indefinitely via retweets and shares, meant that the need for specific items was often addressed before messages requesting them were received via social media channels [45]. Yet despite these limitations, 95% of people surveyed in the affected areas of Tohoku and Kanto expressed their support for government use of social media for the provision of information in disaster situations, and in the remainder of this case study we will focus on the approach taken by government agencies in using social media during the events of March 2011.

4.5.3   Official Use of Social Media

Studies of the Japanese government's use of social media during the Tohoku earthquake and subsequent events reveal that social media, and Twitter in particular, were used extensively throughout this period to communicate information to citizens, at both a central and local government level. Use of Twitter increased dramatically following the earthquake, with over 104 official accounts being held by governmental bodies and public authorities, perhaps the notable being @Kantei_ Saigai, the account of the Prime Minister's Office. [45]. However, although they clearly recognise the value of social media as a communication tool, the Japanese government's use of social media has been criticised as being ineffective, with attempts to reassure the public leading to widespread ambiguity and confusion about the extent and severity of the incident. In fact, it has been suggested that 'official' crisis communication by the government and mainstream media was marginalised in favour of peer-to-peer interactions [39]. Research into the communication strategies employed has attempted to identify why this might be the case.

Content analysis of tweets sent in the 40 h period following the earthquake reveals the disaster prompted several government organisations to register Twitter accounts to communicate with the public throughout the crisis, disseminating information relating to their areas of responsibility, and more general earthquake-related information. The official accounts were very popular with Twitter users, with some local government accounts increasing their follower numbers by at least ten times, which suggests that citizens had an interest in receiving information about the earthquake from an official source [49]. Acar and Muraki's analysis of Twitter activity in the areas affected by the earthquake reveals that one official government account (@ bosai_kesennuma) was particularly active in disseminating warning information to citizens, issuing messages such as:

We've got information that the second wave is bigger than the first. Escape immediately.

However, Twitter activity does not necessarily correspond directly to Twitter *influence*, and analysis of retweeting patterns of official messages reveals that while the number of messages represented an increasing proportion of the sample as events progressed, tweets from government sources were gradually retweeted less than those from 'ordinary citizens' during this time. In Twitter terms this pattern could be interpreted as a loss of influence, as by increasing the number of messages 'the government's aim was to get more information to the public, but this did not translate into information being distributed more widely' ([50], p. 84). A similar analysis of URL links circulated via Twitter shows that the most widely disseminated information-providing links were from peer-generated sources, rather from official government accounts [39].

Although the central government's Twitter accounts had large numbers of followers, thus having the potential to reach a wide audience, their failure to go beyond the straightforward transmission of information resulted in citizens shifting their attention to unofficial channels to fill gaps in their knowledge and situational understanding. This shift in the 'locus of crisis communication leadership' could be attributed to the more 'passive' one-way approach that officials took towards the distribution of information via these accounts (which had only been set up in the aftermath of the disaster), compared with the more interactive peer-to-peer exchanges which took place between members of the public [39]. The lack of influence was exacerbated by a perceived lack of timeliness of government communications [47], and reduced trust due to inconsistencies in the official announcements, particularly in the wake of the incidents at Fukushima nuclear power plant [45]. This perception was reflected in mainstream media coverage, which tended to focus on the relief efforts of local communities and individuals, pointing to a lack of government leadership [47].

Cho et al. (ibid.) argue that these problems are not unique to the Japanese government, and that governments worldwide should reconsider their approach to the use of social media in crisis situations, moving from a linear model to a more networked one to take account of changing communication patterns and retain their credibility.

While Japanese central government has received criticism for being slow to engage with citizens via social media, and failing to facilitate two-way communication, there are some examples of more successful Twitter use at local government levels. At the time of the Great East Japan Earthquake, the local government offices of the city of Tsukuba had been using their Twitter account @tsukubais to communicate with local citizens for several months, with a total of 2000 followers—this number rose to over 10,000 following the events of March 11. Although the area was not as badly affected as some, there was widespread disruption to the city's power, water supply, transport, and communication infrastructure, and the official Twitter account was used to communicate relevant information to citizens. Operational steps were taken to ensure that tweets were disseminated as efficiently as possible, bypassing the usual government approval procedures, and a team of volunteer translators ensured that all messages were also sent in English, Chinese, and Korean, so that they could be understood by the city's non-Japanese speaking residents [49].

Crucially, the @tsukubais account was not simply used for one-way transmission of information. A small team, led by the account manager, engaged in a two-way dialogue with citizens, answering queries, addressing rumours, promoting volunteer activities, and sharing information provided by people in the community, and thus 'became a major source of providing vital information on lifelines and infrastructure until restorations were made after the disaster' (ibid., p. 33).

4.6   Review of Best Practice and Recommendations

Analysis of the three case studies, including comparative evaluation of approaches taking place in similar circumstances where possible, reveals a number of examples of good social media practice. This allows us to identify shared features amongst approaches which proved effective, as well as those which were less successful, and this thematic analysis is presented as a series of recommendations which are outlined in this concluding section. These recommendations are presented with a caveat, as despite numerous commonalities, it is vitally important to note the need for officials and first responders to be sensitive to the specifics of a particular crisis situation. For example, Falkheimer [30] raises the issue that while on the surface terrorist incidents share many features in common with other forms of crisis (such as natural disasters), they present their own specific challenges with regard to communication, including the fact that 'terrorist acts are organized and executed as publicity events in themselves' (ibid. p. 60), and must therefore be managed in ways which take these factors into account. However, evidence from the three events discussed in this chapter highlights the following examples of best practice:

4.6.1 Recommendation 1: Social Media Use Should Be Planned and Established Well in Advance of a Crisis

The case studies demonstrate the importance of developing an approach to social media which is adequately resourced and fully integrated into existing communication strategies. Examples where crisis management teams had an established online presence and were already confident in using social media to engage citizens (such as Greater Manchester Police, or the city of Tsukuba, Japan) showed that this existing infrastructure was used very effectively in the event of a crisis incident, with the public already used to receiving important information via these channels. Conversely, when accounts were less established (or even brand new), these were not utilised as effectively as they could have been. While social media is increasingly used in crisis situations, and Cohen [51] points to the management of Hurricane Sandy in the USA as evidence of this shift, the extent to which individual emergency managers and agencies use social media currently varies greatly. Recent research in the USA suggests that emergency management is in 'a state of transition' while emergency responders adapt to the increased use of social media in their crisis response strategies [4].

4.6.2 Recommendation 2: Strategies Should Be Used to Maximise the Potential of Public Use of Social Media, While Minimising Potential Risks

The examples discussed highlight numerous ways in which members of the public use social media to share information during crisis events, but crisis managers and officials must find ways to make the best use of this potentially vast resource. The case studies highlight some of the potential difficulties in crisis situations, where at best information gathering and dissemination is not as efficient as it could be (as in the Tohoku earthquake, where information was collated on various unconnected volunteer-run sites). At worst, a failure to adequately engage with social media (either by officials leaving a communication 'vacuum', or by failing to fully consider the consequences of some forms of social media interaction) could result in serious operational problems. This can be observed in the 'boat rescue missions' during the Utoya attacks, and in instances where in similar circumstances (such as sieges in Mumbai in 2008 and Kenya in 2013, and the 2014 shooting of a Canadian soldier on sentry duty on Parliament Hill) vital operational details were disclosed to perpetrators via social media posts, which compromised police efforts to resolve crisis situations [27–29].

4.6.3   Recommendation 3: Social Media Should Be Employed in a Timely Manner

As discussed in the previous recommendation, a lack of prompt response from officials creates a knowledge 'vacuum' which is quickly filled by unofficial peer-to-peer exchange of information. Experiences during the Tohoku earthquake and English riots highlighted the importance for emergency managers to monitor social media activity during a crisis event, contributing direct communications of their own to provide accurate, up-to-date and authoritative information, correct misunderstandings, and quash unsubstantiated rumours. In order to do this effectively, organisations may need to modify official processes regarding approval procedures for communications, or delegate responsibility for social media communications to a dedicated account manager.

4.6.4   Recommendation 4: The Official Social Media Source Must Inspire Public Confidence

Despite the need for the rapid dissemination of information, if this information is inaccurate or contradictory, citizens may lose trust in official sources, as demonstrated during the Tohoku earthquake in particular. It is therefore imperative to balance the requirement for communications to be prompt and up to the minute with the requirement for the information provided to be as accurate as possible. While the increased follower numbers of police and government social media accounts during incidents indicates that the public has a strong interest in receiving their information from an 'official' source, other spokespeople or organisations may also form part of an effective communication strategy. The analysis of social media use during the English riots shows that other trusted sources (such as news sources or celebrity figures) may be able to perform an intermediary role, depending on the intended audience of a message.

4.6.5 Recommendation 5: Crisis Communicators Should Engage in Two-Way Communication

Although social media presents a convenient way to disseminate information widely, the cases discussed show that a more interactive approach is more successful in engaging citizens online and is well received by the public (as evidenced in the response to Greater Manchester Police's handling of the 2011 riots). In addition, such an approach provides a valuable opportunity for officials to stem the flow of misinformation and calm situations before they escalate [23]. Research demonstrates the effectiveness of a custom approach which encourages local engagement and utilises local knowledge and expertise (e.g. [52]). A particularly successful example of this approach can be seen in the Queensland Police Service's use of Facebook and Twitter to disseminate information and coordinate discussion during the area's 2011 floods [41, 42].

4.6.6 Recommendation 6: Social Media Messages Must Adopt an Appropriate Tone

The case study examples show that in situations of heightened tension and emotion members of the public are more likely to respond positively to attempts to engage with them if messages are suitably sensitive to variations in local needs and cultural sensibilities. For example, a comparison of Greater Manchester Police's use of social media with that of the Metropolitan Police Service during instances of public disorder in their respective cities reveals that their more empathetic approach to communication was much more positively received by the public. However, while a friendly and empathetic tone is advised, Rasmussen and Ihlen [36] recommend caution in adopting too informal a tone in official social media communications, and suggest that messages should not undermine the authority of the institution sending them.

4.6.7 Recommendation 7: Social Media Communications Should Be Clear and Unambiguous

Evidence from the case study examples, and others, shows the importance of official communications which are accurate, clearly and consistently written, and avoid the use of specialist terminology. Messages should provide a clear description of the danger, its location, and any required action, using language which is easy for citizens to understand (e.g. [36, 53]). In addition, in areas where large sections of the population speak languages other than the main official language, messages should be translated wherever possible (as seen in the Tsukuba government's communications relating to the Tohoku earthquake). The clarity of social media communication can also be improved by the implementation of systems which make messages more straightforward to track and manage, such as including time/date information on messages so that more recent messages can be distinguished from older ones. The use of designated hash tags can also help to identify relevant information, particularly if these are localised to distinguish messages relating to the situation on the ground from more general national or international discussion of the event in question [54].

4.6.8 Recommendation 8: Access to Social Media and Mobile Internet Must Be Considered

Although all three case studies have highlighted the vast potential of social media for crisis communication, this is only the case if citizens have access to the necessary knowledge and technology. Lack of access may occur for many reasons, such as lack of familiarity with social media or infrastructure breakdown in the event of a natural disaster. Therefore, social media must form part of a wider crisis communication strategy, complementing rather than replacing other methods, and crisis managers must consider the specific needs of their target communities when deciding the most appropriate communication methods (e.g. social media, traditional media, local radio, SMS, etc.). For example, during the Haitian earthquake/tsunami of 2010, local knowledge regarding the older age of residents of the island of Hilo, along with the popularity of the local radio station, led emergency managers to choose radio as one of the main channels for disseminating warning information to the community, alongside regular web updates [55].

References

1. Akhgar, B., Fortune, D., Hayes, R., Guerra, B., & Manso, M. (2013). Social media in crisis events: Open networks and collaboration supporting disaster response and recovery. In *Technologies for Homeland Security (HST), 2013 IEEE International Conference*, IEEE, 12–14 November 2013, Waltham MA, USA (pp. 760–765).
2. Houston, J. B., Hawthorne, J., Perreault, M. F., Park, E. H., Goldstein Hode, M., Halliwell, M. R., et al. (2015). Social media and disasters: A functional framework for social media use in disaster planning, response, and research. *Disasters, 39*(1), 1–22.
3. Boin, A., t'Hart, P., Stern, E., & Sundelius, B. (2005). *The politics of crisis management: Public leadership under pressure.* Cambridge: Cambridge University Press.
4. McCormick, S. (2016). New tools for emergency managers: An assessment of obstacles to use and implementation. *Disasters, 40*(2), 207–225.
5. Van Gorp, A., Pogrebnyakov, N., & Maldonado, E. (2015). Just keep Tweeting: Emergency responders' social media use before and during emergencies. *ECIS 2015 Completed Research Papers* (Paper 191).
6. Defence Science and Technology Laboratory (Dstl). (2012). *Smart tips for category 1 responders in using social media in emergency management.* Porton Down, UK: Dstl.

7. Stern, E., & Sundelius, B. (2002). Crisis Management Europe: An integrated regional research and training program. *International Studies Perspectives, 3*, 71–88.
8. Chebib, N. L., Kassem, S., & Rabia, M. (2011). The reasons social media contributed to the 2011 Egyptian Revolution. *International Journal of Business and Management, 2*(3), 139–162.
9. Jurgenson, N. (2012). When atoms meet bits: Social media, the mobile web and augmented revolution. *Future Internet, 4*, 83–91.
10. Moran, M., & Waddington, D. (2015). Recent riots in the UK and France: Causes and commonalities. *Contention: The Multidisciplinary Journal of Social Protest, 2*(2), 57–73.
11. Ball, J., & Lewis, P. (2011, August 24). Riots database of 2.5m tweets reveals complex picture of interaction. *The Guardian*. Retrieved April 23, 2015, from http://www.theguardian.com/uk/2011/aug/24/riots-database-twitter-interaction
12. BBC. (2011). Social media talks about rioting 'constructive'. Retrieved April 23, 2015, from http://www.bbc.co.uk/news/uk-14657456
13. Halliday, J. (2011, August 8). London Riots: BlackBerry to help police probe Messenger looting 'role'. *The Guardian*. Retrieved April 25, 2015, from http://www.theguardian.com/uk/2011/aug/08/london-riots-blackberry-messenger-looting
14. Baker, S. A. (2012). From the Criminal Crowd to the 'Mediated Crowd': The impact of social media on the 2011 English riots. *Safer Communities, 11*(1), 40–49.
15. Procter, R., Vis, F., & Voss, A. (2013). Reading the riots on Twitter: Methodological innovation for the analysis of big data. *International Journal of Social Research Methodology, 16*(3), 197–214.
16. Procter, R., Crump, J., Karstedt, S., Voss, A., & Cantijoch, M. (2013). Reading the riots: What were the police doing on Twitter? *Policing and Society: An International Journal of Research and Policy, 23*(4), 413–436.
17. Glasgow, K., & Fink, C. (2013) Hashtag lifespan and social networks during the London riots. In A. M. Greenbuerg, W. G. Kennedy, & N. D. Bos (Eds.), *Social computing, behavioural-cultural modeling and prediction, Proceedings from the Sixth International Conference, April 2013, Washington, DC, USA.*
18. Lewis, P., Newburn, T., Ball, J., Procter, R., Vis, F., & Voss, A. (2011). *Reading the riots: Investigating England's summer of disorder*. London: Guardian/LSE. Retrieved April 26, 2015, from http://eprints.lse.ac.uk/46297/1/Reading%20the%20riots(published).pdf
19. Panagiotopoulos, P., Bigdeli, A., & Sams, S. (2014). Citizen-government collaboration on social media: The case of Twitter in the 2011 riots in England. *Government Information Quarterly, 31*(3), 349–357.
20. Tonkin, E., Pfeiffer, H. D., & Tourte, G. (2012). Twitter, information sharing and the London riots. *Bulletin of the American Society for Information Science and Technology, 38*(2), 49–57.
21. Vis, F. (2013). Twitter as a reporting tool for breaking news. *Digital Journalism, 1*(1), 27–47.
22. Her Majesty's Inspectorate of Constabulary (HMIC). (2011). *Policing public order: An overview and review of progress against the recommendations of adapting to protest and nurturing the British Model of Policing*. London: HMIC.
23. Baxter, C., Barratt, P., & Thomson, M. (2015). *Social media and the generation, propagation, and debunking of rumours*. Report on behalf of Department of National Defence, Canada. Ontario: Human Systems Incorporated.
24. Crump, J. (2011) What are the police doing on Twitter? Social media, the police and the public. *Policy and Internet, 3*(4), article 7.
25. Copitch, G., & Fox, C. (2010). Using social media as a means of improving public confidence. *Safer Communities, 9*(2), 42–48.
26. Denef, S., Bayerl, P. S., & Kaptein, N. (2013). *Social media and the police: Tweeting practices of British police forces during the August 2011 riots*. Paper presented at CHI 2013: Changing perspectives, April 27–May 2. Paris, France.
27. Bartlett, J., & Reynolds, L. (2015). *The State of the Art 2015: A literature review of social media intelligence capabilities for counter-terrorism*. London: Demos.

28. Oh, O., Agrawal, M., & Raghav Rao, H. (2013). Community intelligence and social media services: A rumor theoretic analysis of tweets during social crises. *MIS Quarterly, 37*(2), 407–426.
29. Simon, T., Goldberg, A., Aharonson-Daniel, L., Leykin, D., & Adini, B. (2014). Twitter in the cross fire: The use of social media in the Westgate Mall terror attack in Kenya. *PloS One, 9*(8), e104136. doi:10.1371/journal.pone.0104136.
30. Falkheimer, J. (2014). Crisis communication and terrorism: The Norway attacks on 22 July 2011. *Corporate Communications: An International Journal, 19*(1), 52–63.
31. Perng, S-Y, Buscher, M, Wood, L, Halvorsrud, R, Stiso, M, Ramirez, L et al. (2012). Peripheral response: Microblogging during the 22/7/2011 Norway attacks. In *Proceedings of the 9th International ISCRAM Conference*. April 2012, Vancouver, Canada.
32. Kaufmann, M. (2015). Resilience 2.0: Social media and (self-) care during the 2011 Norway attacks. *Media, Culture and Society, 37*, 1–16.
33. Christensen, T., Lægreid, P., & Rykkja, L. (2015). The challenges of coordination in national security management—The case of the terrorist attack in Norway. *International Review of Administrative Sciences, 81*(2), 352–372.
34. Vettenranta, S. (2015). Crisis communication and the Norwegian authorities—22 July and the Chernobyl disaster: Two catastrophes, dissimilar outcomes. *Nordicom Review, 36*(1), 51–64.
35. Kamedo (2012). The bomb attack in Oslo and the shootings at Utoya, 2011. *KAMEDO Report 97*. Oslo: Socialstyrelsen.
36. Rasmussen, J., & Ihlen, Ø. (2015). *Lessons from Norwegian emergency authorities' use of social media* (PRIO Policy Brief 14). Oslo: PRIO.
37. SEEDS Asia. (2011). *Damage/needs assessment in the affected area of the 2011 off the Pacific Coast of Tohoku Earthquake and Tsunami*. SEEDS Asia, OYOU International Corporation and Kyoto University. Retrieved May 8, 2015, from http://reliefweb.int/sites/reliefweb.int/files/resources/Fullreport_0.pdf
38. Branigan, T. (2011, March 13). Earthquake and Tsunami: 'Japan's worst crisis since second world war'. *The Guardian*. Retrieved May 8, 2015, from http://www.theguardian.com/world/2011/mar/13/japan-crisis-worst-since-second-world-war
39. Cho, S., Jung, K., & Park, H. (2013). Social media use during Japan's 2011 Earthquake: How Twitter transforms the locus of crisis communication. *Media International Australia, Incorporating Culture and Policy, 149*, 28–40.
40. Jung, J., & Moro, M. (2014). Multi-level functionality of social media in the aftermath of the Great East Japan Earthquake. *Disasters, 38*(S2), S123–S143.
41. Bird, D., Ling, M., & Haynes, K. (2012). Flooding Facebook: The use of social media during the Queensland and Victorian floods. *Australian Journal of Emergency Management, 27*(1), 27.
42. Bruns, A., & Burgess, J. (2012). Local and global responses to disaster: #eqnz and the Christchurch earthquake. *Disaster and Emergency Management Conference, Conference Proceedings*. AST Management Pty Ltd.
43. Ranghieri, F., & Ishiwatari, M. (Eds.). (2014). *Learning from Megadisasters: Lessons from the Great East Japan Earthquake*. Washington, DC: World Bank Publications.
44. Peary, B., Shaw, R., & Takeuchi, Y. (2012). Utilization of social media in the East Japan Earthquake and Tsunami and its effectiveness. *Journal of Natural Disaster Science, 34*(1), 3–18.
45. Appleby, L. (2013). Connecting the Last Mile: The role of communications in the Great East Japan earthquake. London: Internews Europe. Retrieved May 8, 2015, from http://issuu.com/internews-europe/docs/internewseurope_report_japan_connecting_the_last_m
46. Aizu, I. (2011). The Role of ICT during the disaster: A story of how internet and other information and communication services could or could not help relief operations at the Great East Japan earthquake. An article submitted to the Global Information Society Watch Report 2011, to be published by Association for Progressive Communications. Retrieved April 29, 2015, from http://wsms1.intgovforum.org/sites/default/files/webform/igf_wsp/EarthquakeICT0825.pdf

47. Acar, A., & Muraki, Y. (2011). Twitter for crisis communication: Lessons learned from Japan's tsunami disaster. *International Journal of Web Based Communities, 7*(3), 392–402.
48. Murakami, A., & Nasukawa, T. (2012, April 16–20). Tweeting about the Tsunami? Mining Twitter for information about the Tohoku earthquake and tsunami. *WWW 2012*. Lyon, France (pp 709–710).
49. Kaigo, M. (2012). Social media usage during disasters and social capital: Twitter and the Great East Japan earthquake. *Keio Communication Review, 34*, 19–35.
50. Li, J., Vishwanath, A., & Rao, H. R. (2014). Retweeting the Fukushima Nuclear radiation disaster. *Communications of the ACM, 57*(1), 78–85.
51. Cohen, S. E. (2013, March 7). Sandy marked a shift for social media use in disasters. *Emergency Management*. Retrieved June 6, 2016, from http://www.emergencymgmt.com/disaster/Sandy-Social-Media-Use-in-Disasters.html?page=1
52. St. Denis, L., Hughes, A. and Palen, L. (2012). Trial by Fire: The deployment of trusted digital volunteers in the 2011 Shadow Lake Fire. *Proceedings of the 9th International ISCRAM Conference – Vancouver, Canada, April 2012.*
53. Sutton, J., Spiro, E., Johnson, B., Fitzhugh, S., Gibson, B., & Butts, C. (2014). Warning tweets: Serial transmission of messages during the warning phase of a disaster event. *Information, Communication and Society, 17*(6), 765–787.
54. Lachlan, K., Spence, P., Lin, X., Najarian, K., & Del Greco, M. (2016). Social media and crisis management: CERC, search strategies, and Twitter content. *Computers in Human Behavior, 54*, 647–652.
55. Sutton, J., Hansard, B., & Hewett, P. (2011) Changing channels: communicating tsunami warning information in Hawaii. In *Proceedings of the 3rd International Joint Topical Meeting on Emergency Preparedness and Response, Robotics, and Remote Systems, Knoxville, Tennessee.*

Part II
Technological Design and Development of
ATHENA

Chapter 5
Best Practices in the Design of a Citizen Focused Crisis Management Platform

Simon Andrews

5.1 Introduction

Crisis management is an essential function of authorities and organizations that require communication and information processing skills capable of quickly assessing situations and proving actionable information during a dynamic and often chaotic event (See Chaps. 2, 3, and 4). It is essential that citizens are provided with reliable and timely information to keep them safe. Increasingly, the public are making use of the internet and, in particular, social media as sources of their information during a crisis and it is now widely recognized by crisis management practitioners and researchers that this must play a central role in any modern crisis management platform.

Tools are rapidly being developed to assist in harnessing this public information exchange, allowing organizations to more effectively harvest, analyze, aggregate, assess, and disseminate information to better safeguard the public and to make better decisions regarding the actions and deployment of first responders and valuable resources.

This chapter examines some of the best crisis management practices that have arisen in the advent of social media, and considers the tools and technologies that are proving most useful in an effective platform from which to conduct crisis management.

It is these best practices that have helped inform the development of the ATHENA crisis management system [2, 3], along with extensive consultation with crisis stakeholders and experience gained from conducting live exercises with crisis mangers, first responders, and members of the public. Thus, this chapter presents a

S. Andrews (✉)
CENTRIC, Sheffield Hallam University, Sheffield, UK
e-mail: S.Andrews@shu.ac.uk

© Springer International Publishing AG 2017
B. Akhgar et al. (eds.), *Application of Social Media in Crisis Management*,
Transactions on Computational Science and Computational Intelligence,
DOI 10.1007/978-3-319-52419-1_5

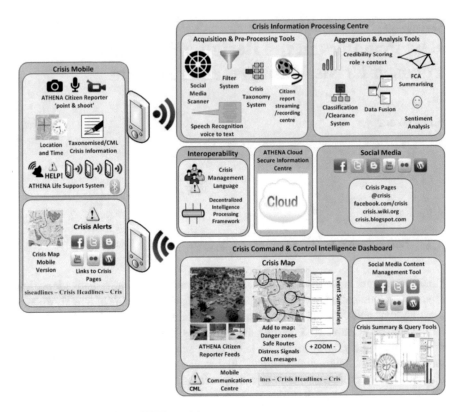

Fig. 5.1 Schematic of the ATHENA crisis management system

review of the current best practice as well as presenting the ATHENA system as an embodiment of beyond the state-of-the-art in best practice for a citizen focused crisis management system.

Figure 5.1 is a schematic view of the ATHENA system. In essence, it is a two-way communication system between the public and crisis managers. It utilizes social media for this communication but also includes a mobile application for citizens and first responders to send eyewitness reports to and receive validated information from a command and control center.

The rest of this chapter is organized as follows: Section 5.2 discusses the use and importance of social media in crisis management. Section 5.3 explores some solutions to managing the problem of information overload in a command and control setting. Section 5.4 examines some approaches to automatically assessing the credibility and priority of information as it comes into the command and control center. Section 5.5 looks at the use of map-based interfaces in crisis management and Sect. 5.6 examines the role that smart mobile devices can play. Section 5.7 details the role that search and rescue application can have during the crisis response phase. Finally, in the light of the previous discussion and from the experiences gained during the ATHENA project and development of the ATHENA crisis management system, Sect. 5.8 lists some recommendations for best practice in designing a citizen focused crisis management platform.

5.2 Social Media in Crisis Situations

Arising originally out of the public's use of social media, local authorities, organizations, and even governments now engage and communicate through the same channels and by the same means. Recognizing how the public were able to self-organize, disseminate information, and help each other through social media, crisis management practitioners began to adopt and adapt accordingly. Initially, organizations, such as Law Enforcement Agencies (LEAs) during the London riots in 2011, realized that social media were key information sources, providing information about activities and events much earlier than their traditional channels of intelligence were able to [7]. Equally, they soon realized that this could be a useful medium with which to disseminate information to and communicate with the public and those involved in the riots. Social media have changed the landscape of public communications and it has been clear for some time that governments and organizations need to follow the trend [10]. During the 2010/2011 Queensland and Victorian floods in Australia, for example, social media played a huge role in safeguarding citizens [4], with individuals and groups self-organizing to obtain and disseminate important information concerning the crisis, in near real-time, and to offer help to those in need of assistance. Similar self-organization and information exchange was also observed during the German floods of 2013 [23]. The public services and local authorities involved in the disasters were also becoming part of the discourse in social media, even to the extent of developing their own social media presences. It was becoming clear that such engagement was useful in order to get messages across to the public and to encourage responses from the public (See Chaps. 4 and 13).

Thus it emerges that the best use of social media in crisis management is as a two-way communication channel crowdsourcing information from the public and feeding information back to the public as a trusted voice in the social media domain. 'An integrated public alert and feedback system that incorporates social media tools that allow for a seamless and straightforward communication from the government to the public and for the public to send relevant information to enhance government operations during a crisis is needed' [9].

This is not to say that social media should supersede current crisis management communication systems or information systems. Social media come with inherent problems of trust and misuse, ethical and legal issues, and the potential for information overload (See Chap. 10). But, 'if leveraged strategically, they can be used to bolster current systems' and improve the effectiveness of existing crisis communication systems [25]. Crisis stakeholders have also recognized the opportunities offered by the use of social media for knowledge exchange between crisis managers and other crisis managers, and between domain experts (such as environmental agencies and utilities providers) and crisis managers [35]. A well thought-out and agreed strategy and format of knowledge exchange and sharing, through social media channels, can assist in achieving a coherent and coordinated crisis response, given the number and diversity (both geographically and in terms of expertise) of organizations and services typically involved in a major crisis.

The European COSMIC Project [11] is investigating the best use of social media in crisis management, and reiterates many of the points made above, including the potential for two-way information flow and for assistance in search and rescue operations (see later), and also suggests that in the aftermath of a disaster, social media can add to society's resilience by providing a common communication medium, a means of organizing and seeking help for rebuilding activities and as a key source of contacts and information concerning relief and regeneration actions and funds [24].

Consequently, social media-based crisis management platforms are now being developed by a range of stakeholders in the field, including researchers, LEAs, and NGOs. For example, The New York City Office of Emergency Management employs the Sahana [30] open-source disaster management software system to manage its network of displaced persons shelters during crisis situations. The news and information channel Al Jazeera, for example, has developed a web-based community service that gathers information from emails, mobile application messages, and text messages to disseminate public sentiment concerning ongoing crisis situations. Ushahidi [33] is a well-known mobile application that allows people to create and send reports of ongoing crisis situations to be displayed on a map of the area. It was used successfully during the Haiti 2010 earthquake to help inform the emergency services of the unfolding disaster [19]. The Dutch government maintains a crisis web site [12] to publish up-to-date information from the government during disasters and emergencies, along with general public safety information to promote crisis preparedness. The site also contains a Twitter feed and a section for FAQs concerning current events.

Information dissemination via social media is clearly an effective and efficient means of pushing reliable information out to citizens during a crisis. However, this can only be the case given sufficient reach and penetration of different social media channels in particular countries, ethnographic areas, and cultures. Consideration also has to be given to the level of technological awareness, age, and language of the desired recipients of the information [9]. Given the preponderance of crises to occur in the less developed areas of the globe, these are important considerations in the development of a crisis management platform appropriate to the target population being safeguarded. For example, the United Nations Office for the Coordination of Humanitarian Affairs (UNOCHA) provides information about humanitarian crises based on information gathered via its reliefweb [32] platform, a gateway to near real-time information on humanitarian emergencies and disasters, the IRIN news service [21] for humanitarian news and analysis, and RedHum [28], a regional humanitarian information network for Latin America and the Caribbean. Each service is tailored to the demographic, geographic, and ethnographic profile of its intended recipients.

Equal consideration, too, is required from the operational perspective: embracing an engagement in social media requires specialized crisis management planning and training. The tools and technologies need to be used with a level of skill, thoughtfulness, and a consideration of ethical and legal issues beyond that of a casual social media user [25]. In particular, the processing and broadcasting of personal information, that may be available in social media, needs to be avoided.

These issues, along with a certain distrust by authorities of information on the Internet, and in social media in particular, may be why the actual take-up and use of social media currently in crisis management systems is quite low. The European EmerGent Project [13] is investigating the current use of social media in crisis management and, in a recent survey of 761 emergency service staff across 32 European countries, found that (a) only a relatively small proportion of staff utilized information from social media during a crisis, (b) most organizations have never actually shared any information with the public during emergencies, and (c) only a small proportion of staff used social media for receiving messages from the public during emergencies [29]. This was despite the fact that the majority of emergency services staff surveyed felt that social media could play a useful role in all of these regards.

The ATHENA Project, therefore, is aiming to bridge this gap between possibility and actuality by developing a system that allows the incorporation of social media as a two-way communication channel and information source within existing systems and operational practice. To achieve this, key stakeholders in crisis management, including emergency services, LEAs, NGOs, and governing authorities, have played a key role in the platform's development, including requirements formation, system testing and validation, and in live exercises, where ATHENA has been deployed in real operational environments, in a number of member states across Europe (see Chaps. 12 and 13 for full details of live exercises).

5.3   Managing Information Overload

The sheer quantity of information produced on the Internet, particularly during a crisis, has meant that comprehensive human monitoring of the plethora of news channels, social media, and web sites that may contain relevant information has become impossible. This has identified the need for the adoption of automated techniques, such as web crawling and web-page scraping, and sophisticated text-based information processing to identify and filter information that may be pertinent to the crisis situation. Social media tools are also required that help crisis managers interpret and analyze information from diverse sources and diverse types, and identify dependencies and relationships between individual pieces of information therein.

The ATHENA system uses an agreed shared vocabulary of crisis situations to harvest crisis-related information from the Internet, in the form of an agreed stakeholder definition of crisis terms and keywords. This is used in the automated identification of relevant web-crawled information and the extraction of crisis-related information, filtering out information that is not recognized as pertinent to the crisis.

Nevertheless, as became apparent during live exercises, even with automated identification and filtering of internet-based information, the large number of individual pieces of crisis-relevant information obtained can still quickly overload the crisis command and control operators. To help mitigate this, a process of automated aggregation, based on text-processing, was implemented to cluster pieces of information

Fig. 5.2 Screenshot of the ATHENA dashboard showing an aggregated crisis report

that were identified as being about the same crisis event [1]. Figure 5.2 is a screenshot of the ATHENA dashboard showing such a cluster. Instead of five individual reports (five separate map-pins) the crisis analyst has a single aggregated report to deal with, although the analyst retains the option to examine each individual piece of information, if desired.

The ATHENA system also provides a range of information filtering options for the crisis manager, including filters for event category, credibility rating, priority rating, and report status (read or unread, validated or not validated). Figure 5.3 shows the ATHENA command and control dashboard in an unfiltered state. Let us say that the crisis manager operating the dashboard is interested in reports of an explosion and only wants to see reports that have not yet been validated. Figure 5.4 shows the same situation, but after filters have been applied for category 'explosion' and report status 'unvalidated,' leaving a single report to focus on.

5.4 Information Credibility

Information is only useful if it is truthful and accurate. Sourcing information from social media and the public brings with it the inherent risk of obtaining contradictory, inaccurate, misleading information. There is always the risk of individuals carrying out a hoax, or deliberately providing malicious or misinformation. This risk is amplified if the crisis involves public disorder or a terrorist attack, where the perpetrators of the crisis are likely to use the very same channels of communication

Fig. 5.3 Screenshot of the ATHENA dashboard in an unfiltered state

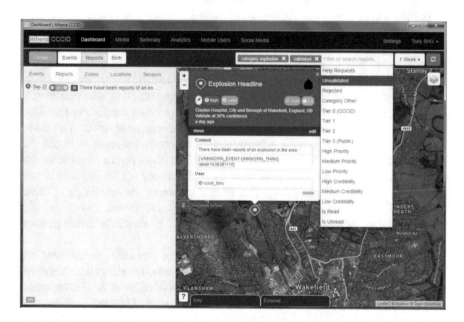

Fig. 5.4 Screenshot of the ATHENA dashboard after 'explosion' and 'unvalidated' filters have been applied

intended for safeguarding to do the very opposite. A key crisis response task is therefore in determining and identifying truths, half-truths, and confusing or contradictory information [14, 18]. Although crisis managers and analysts are expert in such assessments, they cannot be expected to deal with huge volumes of information without the assistance of automation. Dealing with large volumes of 'raw' text-based data obtained from social media feeds 'will usually require some advanced forms of filtering and verification by both machine-based algorithms and human experts before becoming reliable enough for use in decision-making tasks' [6].

The issue of automatically assessing credibility in social media, particularly in Twitter, has been studied in some depth, and various approaches have been devised intended to help ascertain the credibility of text-based information, such as tweets. Many of the approaches have involved so-called machine learning. A test sample of information items, such as tweets, are manually examined by human experts and classified as credible or not credible. The features of each of the classes are determined by inspection, listed and then used in an automated process to classify new tweets. Such an approach is detailed in [8] and in [17] a similar approach has been tailored to 'high impact events.' In a study of 35 million tweets obtained during a number of crisis events, a list of credibility features was extracted, including the number of characters and unique characters present, the use of hashtags, the number of @ mentions, the number of URLs, the presence of swear words, the inclusion of pronouns, and the presence of emoticons. Each feature, its presence, absence and quantity, contribute to a positive or negative credibility score.

In [34], a system that has similarities to Athena is described, in that crisis reports are posted on a shared map by eyewitnesses using a mobile application. The issue of credibility is addressed not only by taking the machine learning approach but also by considering the trustworthiness of the message sender. This was determined by a sender's membership of groups where the groups are already known by crisis managers for their level of trustworthiness. The idea of crowdsourcing was also used, in that users of the system could flag a particular post as being credible or not. Thus posts would acquire a score, or ranking, akin to crowdsourced reviews on retail and holiday web sites. However, the effectiveness of this crowdsourcing approach was unclear and, for ATHENA, it was decided that it was also open to possible misuse and required a level of engagement with users that could not be guaranteed in a crisis situation. Nevertheless, the notion of crowdsourcing for credibility is attractive and ATHENA utilizes it in the way it aggregates information that contain the same event and the same place, thus providing a level of crowdsourced corroboration for that information.

ATHENA uses the machine learning approach for credibility assessment and also uses the trustworthiness of information-senders, although rather than maintaining lists of membership of groups to determine trustworthiness, ATHENA uses a simpler approach of maintaining a list of trusted users. This list includes first responders, people in authority and designated "trusted" roles, as well as citizens who have or develop a track record of supplying reliable information.

In ATHENA, part of the crisis manager's role is to make a final assessment as to whether information presented to them is not only credible but also appropriate and useful to disseminate to the public. The decision to make information public should be based on its usefulness in safeguarding citizens and it should be reliable, unambiguous, timely and clear and have a clear relevance to the public good [18]: 'The public relations literature focuses predominantly on one communication step (dissemination) and one stakeholder group (the public). The results of this study suggest that other essential crisis response communication steps (observation, interpretation, and choice) must be performed as precursors to dissemination and will influence the ability of an organization to rapidly end the response phase and end the risk of immediate damage.' The ATHENA system also allows crisis managers to create their own messages and reports to post, perhaps based on an assessment of incoming information as given above and involving their own expert interpretation or summary of the current situation or ongoing events. 'A [common] misconception about crisis communication is that it is inherently about providing scripted messages designed in advance. We find that crisis communicators would do well to devote more attention to listening to stakeholders involved in the crisis, than in writing anticipated responses that have little chance of contributing to public safety' [31].

During the ATHENA project, during stakeholder consultations and in postexercise briefings, it was clear that command and control staff, who were used to receiving and assessing critical information on a daily basis, trusted information from their own operatives in the field far above that of any other information sources. By categorizing users in terms of levels of trust, staff were thus able to filter information accordingly, focusing, if desired, on reports from their operatives, when they needed to be confident about the credibility of incoming information (See Chap. 13).

Taking this one step further, it was also clear during ATHENA that some organizations wanted a system that safeguarded their staff during crisis situations, but did not want the wider public or internet-based information in the loop. NGOs, for example, with small groups of operatives working in potentially dangerous places, were interested in having an 'ATHENA-lite' system, with a mobile application for their field operatives and a small-scale, laptop-based, command and control dashboard for a senior operative.

5.4.1 Prioritization

Linked to the issue of credibility is the priority of information in terms of taking action and allocating time and resources to respond to it. Clearly, this is an important role of crisis managers, expertly assessing and triaging information as it comes in, but if the volume of incoming information is large, it may be difficult to assess the priorities quickly enough. In a study of crisis management during 15 real crisis situations [18], it was found that there were '[Poor communication filters] in 11 of the 15 primary cases examined. Effective communication filters prioritize information relevant to crisis choice, determining metaphorically, which bells are important and which ones can be safely ignored. However, in these crises, filtering mechanisms failed to help the crisis decision makers handle the deluge of incoming information occurring.' A process of automated prioritization may help to alleviate this problem, when the volume of information is too great, by allowing the user to filter out information unless it has been assessed as 'high-priority.' In ATHENA, alongside the various filtering mechanisms, automated priority assessment is also provided. It is treated as extension of credibility assessment, where priority equals credibility plus severity of event. The severity is determined by text-processing, looking for crisis-related keywords or phrases within the information text, such as citizens being trapped, injured, in danger, drowning, etc. Reports that are automatically assessed as being credible and high priority are highlighted to the crisis manager. A further refinement will be to add a temporal aspect to the analysis it may be considered a lower priority to assign resources to someone drowning if the incident occurred several days ago. Distance between available resources and the event may also be a factor sometimes, when several incidents are requiring urgent attention, and hard decisions need to be made with limited resources where time taken to arrive at the incident may be the deciding factor. However, the current map-based visualization can already provide the crisis manager with the relevant information in this regard.

5.5 Map-Based Interfaces

Disasters and emergencies are usually located somewhere, geographically speaking, thus map-based interfaces for displaying crisis-related information are often used as part of crisis management system, both for command and control and for public facing purposes. SensePlace2 [16], for example, is a map-based web application developed by the Penn State GeoVISTA Center that 'forages place-time-attribute information from the Twitterverse' and uses sense-making tools to provide emergency services and NGOs with a geovisual view of an unfolding crisis. Ushadi [33] is also a map-based application, with geo-located reports sent via mobile devices posted onto a map-based interface. HealthMap [5], developed at Boston Children's Hospital in 2006, is another well-known map-based application, displaying geographically located information concerning outbreaks of diseases and other health-related problems.

ATHENA, too, uses a map-based interface: a PC version for command and control and a mobile version for the public and first responders. ATHENA combines mobile application reports sent in with information crawled from social media, news channels, and other internet-based sources, to produce 'map-pins' attached to relevant, crisis-related, information. Information concerning the same event at the same location is automatically aggregated (forming a single map-pin) to reduce information overload [1] and assist in credibility assessment by providing a level of corroboration.

The interface was tested with real crisis management personnel, first responders, and citizens, during a number of live exercises (See Chaps. 12 and 13). Although a map was found to be appropriate for many of the sense-making and decision-mak-

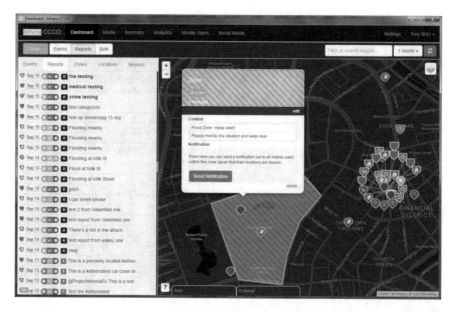

Fig. 5.5 Screenshot of the ATHENA dual map-pin (*right*) and list (*left*) views of crisis information

ing tasks carried out, users often found the dispersed map-pins difficult to navigate and search. The map was useful when a particular location was being examined, but less so when the user was interested in other attributes, such as event type or event time, or when they were carrying out a free-text search. It transpired that an 'email-like' list of the crisis reports was often more useful, with each map-pin corresponding to an item in the list. The list was easier to read through and navigate, and easier to filter, search, and sort. In the end, providing both map-based and email-like views of the same information gave the user the most flexibility and choice. Each operation the user carried out, such as filtering or searching, was mirrored in both views to maintain a consistent display and allow the user to switch easily between the two.

Figure 5.5 is a screenshot of the ATHENA dashboard interface showing the dual map-pin and list views of crisis information. Figure 5.6 shows the similar views but in the mobile application version.

5.6   Mobile Devices and Applications

The widespread ownership of smart mobile devices means that crisis communication and the dissemination of information during a crisis have reached new levels in terms of ease, speed, and quantity of information sharing. Mobile devices have turned citizens into "citizen reporters" with the ability to broadcast eyewitness reports in real time and with a wide range of multimedia content. Mobile internet enabled devices

Fig. 5.6 Screenshots of dual views in the ATHENA mobile application

allow citizens to access news channels and social media instantly and wherever they are. Mobile applications, too, have become ubiquitous and the public are well versed in using them (See Chap. 7). Hence, mobile applications for crisis management and crisis communications purposes are an obvious addition to a crisis management system. The Health Map [5], for example, has a mobile application giving users a localized view of disease outbreak information and alerts of any new outbreaks in their area. A system developed by [34] includes a mobile application with which users can send and share eyewitness reports. Missionmode are an IT company offering crisis management systems including a mobile application for team members to send and receive emergency notifications [27]. Missionmode also list a 'roundup' of 15 other disaster, emergency, and crisis management mobile applications on their web site, most of which include some form of map-based interface to display information. Some of the applications listed are specific to types of crisis such as hurricanes, floods, hazardous substances, or disease outbreaks, whereas others are designed for supplying the public with general safety and crisis preparedness information or designed specifically for first responders or other crisis professionals. One is specifically for finding Red Cross shelters and one is a radio scanner designed to pick up emergency disaster reports from radio stations. About half are cited as being 'global,'

Fig. 5.7 The EarShot mobile app allows sending and receiving of eyewitness crisis reports

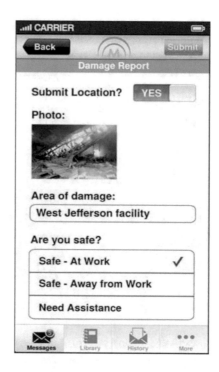

in terms of their geographical coverage, and half "USA" only. Missionmode's own EarShot application is designed for use by first responders and includes the capability of sending eyewitness reports (see Fig. 5.7).

In ATHENA, the aim was to include the public, as well as first responders and command and control, in two-way communication, thus the ATHENA mobile application provides similar eyewitness reporting to EarShot, but aimed at citizens as well as first responders, as in [34]. However, [34] is a 'citizen-to-crowd' application, whereas in ATHENA, command and control are included in the loop as a moderating, trusted voice. Figure 5.8 shows several screenshots showing the "send report" features of the ATHENA mobile application (See Chap. 7).

5.7   Search and Rescue

The locatability of smart mobile devices, through GPS, and their ability to send emergency messages make them very attractive from a search and rescue (SAR) perspective. An important issue, however, is that smart mobile devices are not normally detectable (for good reason, from a privacy standpoint). Rather, they are required to actively send location data to reveal their whereabouts. The SARApp system, for example, facilitates SAR by passing tracking data and geo-tagged evidence from mobile devices to a web-based server for 'real-time situational

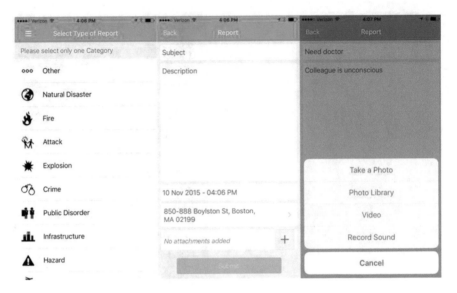

Fig. 5.8 Screenshots of the 'send report' features of the ATHENA mobile application

awareness and analysis' [26]. So long as the application is running, it will continuously broadcast its location. The SARLOC mobile application [20] also queries a
mobile device for its location, but does not require a mobile application. Instead, the
person in trouble is required to contact the SAR team via their mobile device, who
then send a text message to the device, with a link to a web page. Clicking on this
link opens a page in the device's browser which queries the device to identify its
location as a latitude/longitude coordinate. This location data is then automatically
sent to the SARLOC controller and made available to the SAR team. However, there
appears to be few other mature or widespread SAR mobile applications available.

In a crisis situation, it is quite possible that the person requiring assistance is
unconscious or incapacitated in some way and thus unable to operate their mobile
device. A continuously operating application such as SARApp will, of course, send
location data irrespective of the situation of the user, but this does not inform the
SAR team that the user requires assistance. The user is required to notify the emergency services that they are in trouble, or the emergency services need to be
informed some other way, say by a missing person report. The European ISAR+
project [22] is also investigating the issue by looking at methods which do not
require a smart device to broadcast its location, focusing instead on analyzing text-
based information, such as tweets, sent by the person they are trying to locate. Given
that geo-tagged text-data are rare, the project is developing text analysis approaches
to finding 'hints about people's location' from within the text itself [15].

Taking these issues into consideration, and in consultation with SAR stakeholders
in the project, the ATHENA approach is to include a simple emergency 'help' button
in the mobile application. Pressing the button will automatically send location data
to command and control, along with pertinent personal information (such as age,
blood type, and underlying medical conditions) and an optional text message.

5.8 Concluding Recommendations

In the light of the previous discussion and from the experiences gained during the ATHENA project and development of the ATHENA crisis management system, below are listed some recommendations for best practice in designing a crisis management platform (Also see Chap. 4).

1. *Social media*: A crisis management platform should incorporate social media, ideally as a two-way channel of communication and information flow between the public and crisis managers, with crisis managers providing a trusted voice. However, authorities' distrust in these types of information source and the inherent problem of misuse and legal/privacy issues need to be taken into consideration. Social media should not be relied on as the sole/main information source. Designers should also take ethnographic and demographic features of the target population into consideration.

2. *Information overload*: Automated functionality needs to be provided to deal with information overload for crisis managers (and, ideally, for the public, too). Such functionality could include filtering and searching facilities, automatic credibility and priority assessment, extraction of crisis-related information and data from text-based sources, automated classification and categorization of information items, and clustering of similar information items. These functions can assist greatly in providing the user with a clear picture of the unfolding crisis in near real-time.

3. *Information credibility and priority*: Measures should be put in place to assist with the assessment of credibility and priority of information. Tools and techniques exist and are being developed to automatically assess text-based information, such as tweets, for credibility and priority. Aggregating reports that corroborate the occurrence of an event can provide a higher level of credibility than relying on individual reports. Obtaining information from multiple channels/types of source (e.g., social media, news, mobile application) can also provide corroboration. Providing the public, first responders, and other "trusted users" with a mobile application with which to send in eyewitness reports can also assist in obtaining more credible information.

4. *Map-based interfaces*: A map-based interface in which to display crisis events and information is obvious and sensible, but not always appropriate for certain information needs, such as situations where location is not important. Consider providing an alternative, text-based view of the information, such as an 'email-like' list, that is able to be easily filtered, searched, and ordered by the user.

5. *Mobile devices/apps*: A mobile application is a useful addition to a crisis management system. The facility for users to send eyewitness reports can provide useful information, and, in the hands of trusted users such as first responders, information that can be automatically trusted. A mobile application can also act as a receiving tool for safety information disseminated by command and control

and can also provide additional functionality, such as sending distress signals and GPS data to assist in search and rescue actions.

The European COSMIC Project also set out to provide crisis management guidelines and these are given in detail on their web site [11]. However, the focus for COSMIC was on the use of social media whereas here we have taken a slightly broader view, encompassing other elements of a modern crisis management platform. It is hoped that crisis managers and developers of crisis management platforms can treat these sets of recommendations as complementary and find them useful in designing systems that provide effective and efficient tools with which to help safeguard citizens in crisis situations.

References

1. Andrews, S., Gibson, H., Domdouzis, K., & Akhgar, B. (2016). Creating corroborated crisis reports from social media data through formal concept analysis. *Journal of Intelligent Information Systems, 47*, 287–312.
2. Andrews, S., Yates, S., Akhgar, B., & Fortune, D. (2013). The ATHENA Project: Using formal concept analysis to facilitate the actions of responders in a crisis situation. *Strategic intelligence management: National security imperatives and information and communication technologies* (pp. 167–180). Elsevier: Butterworth-Heinemann.
3. ATHENA. (2016). "The European ATHENA Project—Use of new smart devices and social media in crisis situations. Retrieved September, 2016, from http://www.projectathena.eu/
4. Bird, D., Ling, M., & Haynes, K. (2012). Flooding Facebook—The use of social media during the Queensland and Victorian floods. *The Australian Journal of Emergency Management, 27*(1), 27.
5. Boston Children's Hospital: HealthMap. (2016). Retrieved September, 2016, from http://www.healthmap.org/en/
6. Boulos, M. N. K., Resch, B., Crowley, D. N., Breslin, J. G., Sohn, G., Burtner, R., et al. (2011). Crowdsourcing, citizen sensing and sensor web technologies for public and environmental health surveillance and crisis management: Trends, OGC standards and application examples. *International Journal of Health Geographics, 10*, 67.
7. Briggs, D., & Baker, S. A. (2012). From the criminal crowd to the 'mediated crowd': The impact of social media on the 2011 English riots. *Safer Communities, 11*(1), 40–49.
8. Castillo, C., Mendoza, M., & Poblete, B. (2011). Information credibility on twitter. In *Proceedings of the 20th International Conference on World Wide Web* (pp. 675–684). ACM.
9. Chan, J. C. (2010). *The role of social media in crisis preparedness, response and recovery* (Vanguard).
10. Cole, R. (2009). Social media: What does it mean for public managers? *Public Management, 91*(9), 8–12.
11. COSMIC. (2014). The European COSMIC Project—Contribution of social media in crisis management. Retrieved September, 2016, from http://www.cosmic-project.eu/
12. Dutch Ministry of Security and Justice. (2016). Retrieved September, 2016, from http://www.crisis.nl/
13. EmerGent. (2014). *The European EmerGent Project—Emergency management in the social media generation.* Retrieved September, 2016, from http://www.fp7-emergent.eu/
14. Engemann, K. J., & Miller, H. E. (1992). Operations risk management at a major bank. *Interfaces, 22*(6), 140–149.

15. Flizikowski, A., Przybyszewski, M., Stachowicz, A., Olejniczak, T., & Renk, R. (2015). Text analysis tool tweet locator–TAT2. In *Proceedings of the 12th International ISCRAM Conference*. Kristiansand, Norway.
16. GeoVISTA. (2011). *SensePlace2*. Retrieved September, 2016, from http://www.geovista.psu.edu/SensePlace2/
17. Gupta, A., & Kumaraguru, P. (2012). Credibility ranking of tweets during high impact events. In *Proceedings of the 1st Workshop on Privacy and Security in Online Social Media* (p. 2). ACM.
18. Hale, J. E., Dulek, R. E., & Hale, D. P. (2005). Crisis response communication challenges building theory from qualitative data. *Journal of Business Communication, 42*(2), 112–134.
19. Heinzelman, J., & Waters, C. (2010) *Crowdsourcing crisis information in disaster-affected Haiti*. US Institute of Peace.
20. Hore, R. (2012). *SARLOC mountain rescue application*. Retrieved September, 2016, from http://www.go4awalk.com/the-bunkhouse/walking-news-and-discussions/walking-news-and-discussions.php?news=710222
21. IRIN. (2016). *The inside story on emergencies*. Retrieved September, 2016, from http://www.irinnews.org/
22. ISAR+. (2015). The *European ISAR+ Project—Online and mobile communications for crisis response and search and rescue*. Retrieved September, 2016, from http://isar.i112.eu/index.html
23. Kaufhold, M. A., & Reuter, C. (2016). The self-organization of digital volunteers across social media: The case of the 2013 European floods in Germany. *Journal of Homeland Security and Emergency Management, 13*(1), 137–166.
24. Kotsiopoulos, I. (2014). Social media in crisis management: Role, potential, and risk. In: Utility and Cloud Computing (UCC). *2014 IEEE/ACM 7th International Conference* (pp. 681–686). IEEE.
25. Merchant, R. M., Elmer, S., & Lurie, N. (2011). Integrating social media into emergency-preparedness efforts. *The New England Journal of Medicine, 365*, 289–291.
26. Metron. (2014). *SARApp, search and rescue application*. Retrieved September, 2016, from http://www.sarapp.com/
27. MissionMode. (2016). *Mobile control*. Retrieved September 26, 2016, from http://www.missionmode.com/solutions/-module/mobile-control/
28. Redhum. (2016). *Humanitarian information network for Latin America and the Caribbean*. Retrieved September, 2016, from http://www.redhum.org/
29. Reuter, C., Ludwig, T., Kaufhold, M. A., & Spielhofer, T. (2016). Emergency services attitudes towards social media: A quantitative and qualitative survey across Europe. *International Journal of Human-Computer Studies, 95*, 96–111.
30. Sahana Foundation. (2016). *Open source disaster management software*. Retrieved September, 2016, https://sahanafoundation.org/
31. Ulmer, R. R., Sellnow, T. L., & Seeger, M. W. (2013). *Effective crisis communication: Moving from crisis to opportunity*. Thousand Oaks, CA: Sage.
32. UN Office for the Coordination of Humanitarian Affairs (OCHA). (2016). *Reliefweb*. Retrieved September, 2016, http://reliefweb.int/
33. Ushahidi. (2016). *Read the crowd*. Retrieved September, 2016, from https://www.ushahidi.com/
34. Weaver, A. C., Boyle, J. P., & Besaleva, L. I. (2012). Applications and trust issues when crowd-sourcing a crisis. In *2012 21st International Conference on Computer Communications and Networks (ICCCN)* (pp. 1–5). IEEE.
35. White, C. M. (2011). *Social media, crisis communication, and emergency management: Leveraging Web 2.0 technologies*. New York: CRC Press.

Chapter 6
Analyzing Crowd-Sourced Information and Social Media for Crisis Management

Simon Andrews, Tony Day, Konstantinos Domdouzis, Laurence Hirsch, Raluca Lefticaru, and Constantinos Orphanides

6.1 Introduction

Social media is rapidly changing the way emergency data is created and distributed during a crisis. The proliferation of mobile devices together with ubiquitous tools for online dissemination means that the growing pool of dynamic changing and updating information is of increasing importance to emergency personnel in a crisis situation (see Chaps. 2–4). Crowdsourcing can provide the fastest access to localized information and a number of studies have suggested that emergency response can be improved by employing local community knowledge that then allows them to provide aggregated situational awareness emerging in real time [1, 2].

However, the challenges of using such data have also been highlighted and, indeed, it has been recognized that social media data can also be a source of misinformation, propaganda, and rumor, both intentional and unintentional [3, 4]. The obvious risks associated with using an unregulated stream of information imply that assessing the reliability of crowd-sourced data has emerged as a crucial task [4]. Gao [5] also points out that the level of messaging in a disaster will be so high that meaningful filtering, aggregation, categorization, and summarization will be essential capabilities if we are to make effective use of the data in a crisis response capability. Narvaez [6] further argues that appropriate organization of social network information is the key to providing support in terms of ground action.

The ATHENA project aims to harness available data from a variety of sources as a way of extracting actionable intelligence for the public and first responders. This is achieved by developing systems that assist in the searching, acquisition, aggregation, filtering, and presentation of knowledge from social media and crowd-sourced

S. Andrews (✉) • T. Day • K. Domdouzis • L. Hirsch • R. Lefticaru • C. Orphanides
CENTRIC, Sheffield Hallam University, Sheffield, UK
e-mail: S.Andrews@shu.ac.uk; T.Day@shu.ac.uk; K.Domdouzis@shu.ac.uk;
L.Hirsch@shu.ac.uk; R.Lefticaru@shu.ac.uk; C.Orphanides@shu.ac.uk

© Springer International Publishing AG 2017
B. Akhgar et al. (eds.), *Application of Social Media in Crisis Management*,
Transactions on Computational Science and Computational Intelligence,
DOI 10.1007/978-3-319-52419-1_6

data to support crisis management. In this chapter, we describe the technology employed at each of these stages which supports the challenges associated with the fact that the raw data used in these processes comes from a variety of sources and is often difficult to process and analyze, as it is by its nature loosely structured and noisy. Real examples are useful in development and evaluation of the tools and also in explaining how a system works. We therefore refer to a particular data set, namely a set of tweets transmitted during the Colorado wildfires in 2012 [7], to demonstrate aspects of the system, such as development of the crisis taxonomy or delivering credibility assessment.

We include explanation of how we use and combine various commercial and freely available open-source software tools to support the tasks. We also describe a powerful visual instrument, enabled by formal concept analysis, which we have developed. The tool allows an analyst to drill down into the concept hierarchy which has been established in previous stages.

The chapter proceeds as follows. In Sect. 6.2, we begin by describing how we obtain and structure the data. We include an explanation of the information acquisition process and the data processing pipeline that has been established. Assessing the credibility and priority of any crisis information is obviously a critical task, and in Sect. 6.3 we discuss the challenges involved and the process employed in ATHENA, together with informative examples from the Twitter data. In recent years, the research community has been very active in the area of sentiment analysis and it is clear that establishing the sentiment and any changes in sentiment from large volumes of crowd-sourced information can be extremely useful to analysts and first responders, and in Sect. 6.4 we discuss how we track and report sentiment information and include a discussion around the supporting taxonomy. In Sect. 6.5, we review the use of formal concept analysis in the data aggregation process and in Sect. 6.6 we discuss filtering and searching.

6.2   Obtaining Structured Crisis Data

The ATHENA project has developed data collection and preprocessing tools for the real-time acquisition of data, such as text, images, video, and voice, from social media and from mobile devices. The ATHENA project is characterized by an information processing process that must collect and provide robust, high quality data to the rest of the components. This process is based on two stages. The first stage is the acquisition and preprocessing of data. The second stage is based on data analysis and aggregation of the collected data. A further component, the crisis command and control intelligence dashboard (CCCID), is responsible for the presentation of this data to the user.

In ATHENA, information acquisition is the process of obtaining crisis data from a number of sources including social media websites, such as Facebook and Twitter, as well as reports sent directly to the system via the ATHENA mobile application. ATHENA will then use a number of filtering processes and crisis taxonomies in order to focus on information that can be formulated to valuable intelligence.

Social media is considered a valuable source of crisis data. Twitter, in particular, is often used to share and disseminate crisis information. A tweet can include a variety of information, such as text, images, videos, audio, and additional links. In addition, there is also a significant amount of metadata that is attached to each tweet. This metadata includes information such geolocation (either a place or geo-coordinates), the author name and Twitter handle, a defined location, a timestamp of the moment the tweet was sent, the number of retweets, the number of favorites, a list of hashtags, a list of links and other users mentioned in the tweet. All of this data can be valuable when attempting to extract information for use in crisis response. Facebook also provides access to the data hosted on it; however, access is much more limited. Therefore, ATHENA mainly proposes to utilize content posted directly to the ATHENA Facebook page or, potentially, other relevant crisis pages extracting both posts and comments.

Given that much of the content shared on social media often links to external websites, ATHENA will also look to analyze the content of the pages these links point to. This content may be news reports containing standard information but in other cases, these may contain live reports posted by journalists that are not detected via the Twitter and Facebook crawlers or posts by individual bloggers.

ATHENA also collects crisis data through the ATHENA mobile application (see Chap. 7). The app is customized for different users: public users who have access only to publicly available information and two tiers of trusted users. The highest level of trusted users would be members of the police, other blue light services, and first responders; and the lower level of trusted users would include official volunteers in the emergency response, public officials, and members of utility companies. Having a trusted status within the app means that the user is able to view information not made available publically. Data received through the mobile app may include a category text, images, video, audio, geolocation, a timestamp, and the status of the user who submitted the report.

Currently, data from social media and the web is acquired using SAS Information Retrieval Studio (IRS) which provides specific social media crawlers, web crawlers and the ability to monitor RSS feeds. IRS provides a data processing pipeline that connects each SAS component together. It involves five main stages. The first stage is the data import stage as described in this section. This stage is related to the instantiation of the social media crawlers for ATHENA by inputting relevant search terms to monitor specific data (Fig. 6.1). The second stage is filtering. In this case, any incoming information is evaluated for its relevance to the crisis based on certain criteria which may be keywords, categorization, credibility, etc.

The third stage is the categorization, context and contextual extraction phase. This stage identifies keywords, categories, and concepts in each of the posts. The

Fig. 6.1 Overview of the ATHENA information processing pipeline

fourth stage identifies any sentiment expressed in each of the posts while the fifth stage involves the exporting of the data so that it can be utilized by the CCCID.

ATHENA uses the SAS content categorization studio in order to extract categories, concepts or entities, and the identification of relationships between concepts (contextual extraction) from documents. It enables the user to build taxonomies and rules that can then be imported into IRS and applied to the documents as they arrive in real time.

Categorization is the process of analyzing a document's content and applying a category to it based on a set of predefined conditions. In ATHENA, documents can belong to a number of categories if there is match between the content of the documents and the rules from different categories. SAS content categorization studio offers automatic rule generation, rule writing and statistical categorization. We utilized the rule writing process whereby categories are built by constructing a set of Boolean rules that is applied to the incoming documents. By connecting multiple Boolean rules, rules can be made more complex.

A crisis categorization taxonomy has been developed in order to categorize each collected post into one or more categories and is shown below in Fig. 6.2. The list of possible crisis events has been categorized into seven top-level types: Attack, Crash, Hazard, Health, Natural Disaster, Other, Public Order Incidents, Terrorism. Each of these is subdivided to further categories. For example, an attack can be categorized in bomb attacks, hostages, killings, knife attacks, lone wolf attacks, shootings, and suicide bombs.

Concept extraction is used to extract more specific details from posts than categorization does. Such details include locations, people, and organization but may also relate to specific contextual factors (e.g., information specifically related to a crisis). Concept and contextual extraction is the process of identifying these features and entities in the document, as accurately as possible, and extracting them. These features may be related to keywords, particular sentence constructs, or even relationships between these features. An example of the concepts that may be extracted from a document is shown in Fig. 6.3.

Once the documents are categorized and had their concepts extracted, through a process of data booleanization and discretization [8], the data are transformed into formal contexts, making the data accessible by the knowledge discovery and intuitive, conceptual visualization techniques [9] of Formal Concept Analysis (FCA). For an overview of FCA, see Sect. 6.5.

6.3 Assessing Credibility and Priority

According to the Oxford dictionary, credibility is 'the quality of being trusted and believed in, the quality of being convincing or believable.'[1] One objective of ATHENA is to develop tools and techniques 'to automatically assess information

---

[1] https://en.oxforddictionaries.com/definition/credibility.

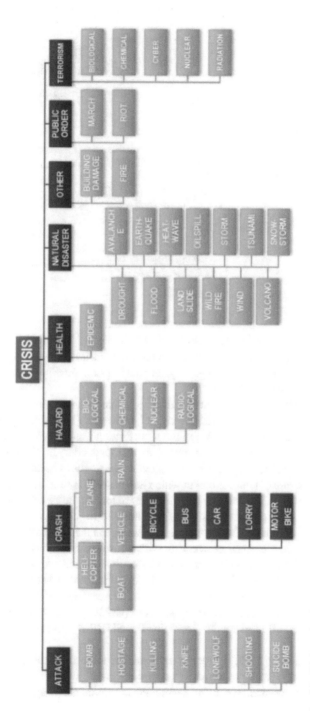

Fig. 6.2 Crisis categorization taxonomy

Fig. 6.3 Example of concepts extracted from a tweet

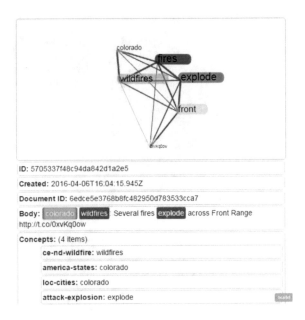

ID: 5705337f48c94da842d1a2e5	
Created: 2016-04-06T16:04:15.945Z	
Document ID: 6edce5e3768b8fc482950d783533cca7	
Body: colorado wildfires Several fires explode across Front Range http://t.co/0xvKq0ow	
Concepts: (4 items)	
	ce-nd-wildfire: wildfires
	america-states: colorado
	loc-cities: colorado
	attack-explosion: explode

and information sources for credibility and maliciousness.' In order to achieve this, and subsequently prioritize the information received, we are focused on two targets: detecting credibility and, consequently, enhancing the provision of situational awareness in a crisis situation.

Within ATHENA we have identified three possible measures on which to calculate the credibility of the information received, these are: (1) who has provided the information, e.g., in the ATHENA app the report's author is known when they are a registered public user, an official volunteer, or first responder; (2) the text itself, on which classification techniques can be applied to assess the content credibility and informativeness; and (3) the context in which the message is sent, including both the content and corroboration level (more messages reporting the same incident, previously reported and already validated, will contribute to increasing the credibility and automatic validation of new incoming messages on the same topic).

In ATHENA we may well be working with information received from multiple sources via different channels and contexts; in this section we will present our approaches to automate the detection of credible and informative crises related messages. As detailed in Sect. 6.2, the information received by ATHENA Crisis Command and Control Intelligence Dashboard (CCCID), previously described in [10], has multiple different sources including the ATHENA app, social media, RSS, and news sites.

We expect the two main sources of information to be from the app and social media (most likely Twitter). In order to assess the credibility of data coming from these two main streams, we have developed complimentary approaches which depend on the information source. Firstly, messages received from the ATHENA app may be: (a) public reports; (b) trusted user tier-2 reports, or (c) trusted user tier-1 reports. Reports from type (b) and (c) have both a higher credibility (as the reports are coming from 'crisis professionals') and a higher priority, since they could present useful details, warnings, or important actions that need to be taken (see Chap. 5).

All the public messages received from the mobile app (which may include anonymous, unregistered users) will be evaluated by human operators in the CCCID and consequently validated or rejected (their initial status is un-validated). A help message (a specific type of app report indicating a user requires urgent assistance) will be validated if it appears credible and genuine, otherwise it can be rejected, if, for example, the message's content indicates the app is being misused, the message is not urgent or relevant, and other contains malicious intentions.

These operators will be assisted by a semi-supervised machine learning algorithm which will suggest if the message should be validated or rejected based on the past history or validated and rejected messages made by the operators. The automatic label proposed by the algorithm will not be definitive and can be modified by the operators. After such a modification the labelled message will be added to the training set of the machine learning algorithm; therefore, the algorithm should be constantly improving its accuracy and when encountering a similar message is the future its prediction should be closer to that of a human operator.

In addition to data received from the app, ATHENA also retrieves messages and posts from social media which have been detected to contain crisis relevant information. For this second case we propose an automatic classification, using supervised machine learning techniques, the methodology to be followed is detailed in Sect. 6.3.1.

On top of assessing the credibility of the received messages, another important task is to automatically prioritize these messages so the users of the CCCID know which messages need attention most urgently. This is especially important as the amount of incoming information may be large and arriving quickly. To the best of our knowledge, this is a new research area and the few existing techniques that have emerged are concerned with crisis situations generated by floods. In these cases, the proposed method is a geographical prioritization of social network messages using sensor data streams [11]. The approach is based on analyzing previous data and observing the correlation between the number of Twitter messages near flood affected areas, which appear in a much higher concentration compared to other areas [11, 12], and suggesting the combination of geographical data from the sensors with social media to prioritize and improve situational awareness during floods.

Currently, the CCCID prioritization mechanism is based solely on the reporting user's tier—with messages from trusted users taking a higher priority. However, this new approach of combining geo-data with social media could be very useful when sensor data streams from the authorities are available and it should be investigated in other cases of natural disasters, e.g., earthquakes.

Table 6.1 Sample of tweets posted during the Colorado fires (2012) and their classification

Tweet text	Informativeness
****** and her husband had to leave their horses and evacuate immediately from the *#HighParkFire*. ***********	Related and informative
Thousands evacuated as *Colorado wildfire* nears http://t.co/1jtGu6Bb	Related and informative
RT @**********: Hey *Colorado Springs* radio stations—Let's go ahead and remove Adele's Set *Fire* to the Rain from our rotation #JustSa	Not related
RT @***********: Lord, protect those in the path of the fires in *Colorado*. Please, send your Divine extinguisher for those *fires*!	Related but not informative

6.3.1 Credibility Assessment of Twitter Messages

Twitter is one of the most popular sources of crisis data due, both to its volume and the fact that it is relatively simple to access. It is also easy for people to report and share crisis-related information through its platform. Thus it is possible that Twitter data could contribute to situational awareness within ATHENA for Law Enforcement Agencies (LEAs), police, and first responders. However, the huge amount of messages transmitted via Twitter or other social media makes it impossible for human operators to manually cull and extract relevant information in an emergency situation. Consequently, Natural Language Processing (NLP) and Machine Learning (ML) have emerged as key techniques for automating the extraction of situational awareness information broadcasted via social media [13–15].

This justifies the development of appropriate classifiers for detecting credible messages and this problem has been studied by many researchers in an off-line (or post-hoc) setting, using data gathered from previous high impact crisis situations to train the classifiers (see Chap. 8).

The majority of researchers have focused on analyzing Twitter data obtained via Twitter API, which is currently easier to access than Facebook, which has stricter rules of access and limits access to posts on public pages. Datasets of tweets transmitted during high impact crises have been previously collected and labelled in order to serve as training and test sets for the classifiers. An example of tweets transmitted during the Colorado fires in 2012 is given in Table 6.1. The keywords which might relate them to the crises are emphasized; however, the personal data, such as name or usernames, has been hidden for ethical reasons. The table contains different tweets which were labelled in the study [7] into different categories or classes such as related and informative; not related; not applicable; related but not informative.

In Fig. 6.4 we present an overview of our research approach, aiming to apply ML (more precisely supervised learning) and NLP techniques in order to build a classifier to assess the credibility and information awareness of social media messages sent during crises situations. The first step, data collection, consists of obtaining the appropriate datasets and labelling them,[2] i.e., annotating each record with the class

---

[2] Several labelled data collections, including the Colorado wildfires dataset used in this chapter, have been downloaded from http://crisislex.org/data-collections.html.

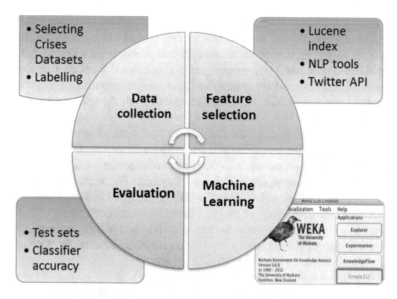

Fig. 6.4 Overview of the machine learning process for credibility assessment

to which it belongs. The collection we considered for training consists of a labelled dataset of 1200 tweets extracted from those sent during the Colorado wildfires (2012), which is presented in [7].

In the ML process, the simple text alone is not enough for training; therefore, for each tweet additional features are extracted or computed, representing attributes which might be relevant for establishing the credibility and message informativeness. The features computed for each tweet could be grouped into different scopes as in [16]:

- Message related: length of the text of the tweet (in characters); number of words; fraction of capital letters in the tweet; number of URLs contained on a tweet; mentions a user (e.g., @username); includes a hashtag (e.g., #HighParkFire)
- User related: registration age (time passed since the author registered their account in days); statuses count (number of tweets at posting time); number of people following this author at posting time; number of friends (number of people this author is following at posting time) if the user has a verified account
- Topic: number of tweets, average length
- Propagation: number of retweets, degree of the root in a propagation tree, etc.

The list above is not exhaustive; there are authors using larger sets, containing for example 45 features [17], in order to determine the credibility of a tweet. There are also studies showing which features are the better indicators of credibility and occur more often in data describing emergency situations [18]. For a detailed survey of existing research and tools developed for processing social media messages in mass emergency [19] can be consulted.

However, for privacy issues and other concerns regarding personal data we did not employ any user-related features (such as number of followers on Twitter), and we focused on the message-related attributes. It is worth mentioning that in the feature selection we have also considered our calculated sentiment of the tweet (detailed in Sect. 6.4) and the number of identified categories (see Fig. 6.2) and crisis-related concepts which have been extracted from the text, according to the taxonomy presented in Fig. 6.3.

After processing the appropriate features for each record in the training dataset, the third stage of the process is the ML stage. Through the use of the ML software WEKA [20], we have experimented with the use of a number of different ML algorithms such as AdaBoost, Naive Bayes, Bayesian Networks, IBk (based on K-Nearest Neighbors), decision trees like J48 (an open-source implementation of the C4.5 algorithm), and decision tables. Currently, we are performing comparisons between these algorithms, in order to determine the ones which are more appropriate for our problem, one particular algorithm providing good performance for the current dataset is J48, an algorithm used to generate a decision tree.

The fourth step consists of evaluating the classifier's ability to predict the correct labels for the testing set. The classifier cannot be evaluated on the same data set which was used to train the classifier; however, in the absence of a clear training and testing set the most common approach is to use a tenfold cross validation strategy. The strategy partitions the data set into ten groups of which nine are used for training and one is used for testing. This approach is repeated ten times each time using a different group as the testing set. The approach we have presented is in the evaluation phase and will be subsequently integrated in the ATHENA data processing pipeline. Currently, the accuracy of the classifier is around 80%; however, it could be improved if other features, such as those which are user related, were taken into account.

6.4 Sentiment Analysis

Sentiment analysis is used to categorize and classify the opinions and sentiments expressed in text. In this case we again situate our explanation in the Colorado wildfire Twitter dataset. We have used SAS sentiment analysis studio to develop our sentiment models and evaluate the polarity of the text overall and polarity with regard to particular entities or features within the text. This polarity can be positive, negative, or neutral. The following polarity classes were implemented:

- Positive—a positive sentiment has been expressed
- Negative—a negative sentiment has been expressed
- Neutral—a neutral sentiment has been expressed
- Unclassified—the sentiment expressed does not fall in any of the defined polarity classes

The polarity of each document is measured at both the overall document level (i.e., the whole tweet), and, when applicable, at the specific feature level (i.e., sentiment

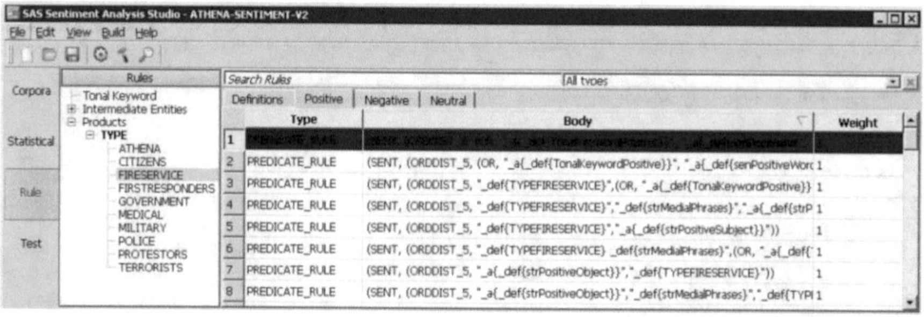

Fig. 6.5 The ATHENA sentiment analysis model (partial screenshot)

explicitly expressed towards an entity involved in the crisis event). This is possible through the creation of multilevel taxonomies to assess sentiment, as explained in the subsection below.

6.4.1   Sentiment Taxonomy

The ATHENA sentiment taxonomy utilizes a rule-based model that uses sentiment vocabularies and handcrafted rules, custom-tailored to crisis events. These rules comprise term matching, regular expressions, and part-of-speech tags, along with pre-built Boolean operators expressing constraints, such as the distance and occurrence of concepts in relation to other words.

Figure 6.5 shows some of the custom-built predicate rules defined for the purposes of ATHENA, which allow for the definition of semantic relationships between concepts (i.e., entities). The left column displays some of the concepts which are usually identified in crisis management situations, such as citizens, first responders, and law enforcement agencies. Each entity can have its own set of rules, allowing for contextual identification of sentiment; the fourth rule, for example, looks for instances of the fire service concept combined with words expressing positive sentiment, in the same sentence, having a maximum distance of five words between the two.

6.4.2   Example of ATHENA Sentiment Analysis

Figure 6.6 shows a partial example of the sentiment identified in the Colorado wildfires Twitter dataset. The overall sentiment of each document (tweet) is expressed in the first column. When sentiment has been identified for specific concepts in each document, it is expressed in the third column. The fourth row contains the actual body of each tweet, which has been purposefully truncated for ethical purposes.

Inspecting the results of the first row, for example, shows how an overall positive sentiment was identified for that particular tweet, but also how the positive sentiment

sentiment	product-sentiment	feature-sentiment	tweet-body
Positive	TYPE--Positive	TYPE--FIRESERVICE--Positive	colorado fire fighte...
Unclassified			Nexus 7
Negative			Hundreds of homes de...
Positive			RT @JoViClo: God Ble...
Negative	TYPE--Negative	TYPE--CITIZENS--Negative	My prayers go out to...
Positive	TYPE--Positive	TYPE--MILITARY--Positive	So proud of the Colo...
Positive	TYPE--Positive	TYPE--FIRESERVICE--Positive	Colorado wildfire ho...
Neutral			RT @themusicninja: H...
Negative	TYPE--Negative	TYPE--FIRESERVICE--Negative	Federal firefighters...
Negative	TYPE--Negative	TYPE--CITIZENS--Negative	RT @kmitchellDP: Col...
Positive			shout out to colorad...
Negative	TYPE--Negative	TYPE--GOVERNMENT--Negative	#newbedon 6/26/2012 ...
Negative	TYPE--Negative	TYPE--CITIZENS--Negative	RT @kariontour: Pray...

Fig. 6.6 An example of identified sentiment in the Colorado wildfires Twitter dataset

was expressed towards the Colorado fire service, which evidently handled the crisis successfully. In fact, out of the 1200 tweet corpus, a majority of the sentiment expressed towards governmental entities such as the fire service and the military has been positive.

6.5 Aggregation to Reduce Information Overload

In a crisis situation, decision makers need a clear picture of the events occurring. It is no use being overloaded with information from hundreds or even thousands of sources, as can be retrieved from social media. Thus the ATHENA system has a process to aggregate sources when they contain information about the same event, greatly reducing the number of information points presented to the decision maker. Furthermore, this aggregation can give an indication of the size, seriousness, and credibility of the event simply by the number of sources involved (although the number of corroborating sources should not, of course, be relied upon as the only measure of these factors). In ATHENA, this aggregation is carried out by a clustering technique called Formal Concept Analysis (FCA) [21].

A formal description of formal concepts begins with a set of objects G and a set of attributes M. A binary relation $I \subseteq G \times M$ is called the formal context. If $i \in G$ and $j \in M$ then iIj says that object i has attribute j. For a set of objects $A \subseteq G$, a derivation operator $'$ is defined to obtain the set of attributes common to the objects in A as follows:

$$A' : \{j \in M | \forall i \in A : iIj\}$$

Similarly, for a set of attributes $B \subseteq M$, the operator is defined to obtain the set of objects common to the attributes in B as follows:

$$B' : \{i \in G | \forall j \in B : iIj\}$$

(A, B) is a formal concept if and only if $A' = B$ and $B' = A$. Thus A and B have the following properties: (1) Every object in A has every attribute in B, (2) For every object in G that is not in A, there is an attribute in B that that object does not have, and (3) For every attribute in M that is not in B, there is an object in A that does not have that attribute.

In ATHENA, the information sources are the objects and the structured data extracted from them are their attributes. Thus an object might have attributes such as a location, a crisis category, a sentiment, a date and time, and so on. If the aggregated information is to be presented to the decision maker via a map of the crisis area, we can define a 'crisis concept' as being a formal concept that contains at least one location and at least one crisis category. Thus FCA has been implemented in ATHENA to compute crisis concepts from the structured data obtained from the social media and citizen reporter information sources.

The Colorado wildfire Twitter data provides an example of this and Fig. 6.7 shows part of a formal concept tree generated from the structured data obtained from the tweets. Each node is a formal concept and the node on the left represents the set of 642 tweets that have a location and a crisis category in their text. Each of the numbered nodes on the right is a 'crisis concept' containing a location and a crisis category. The label above each node contains the attributes of the concept and the label below gives the number of tweets sharing those attributes. If a node is filled in, it means there are further specialized 'sub-concepts' that can be explored—containing fewer tweets but a greater number of shared attributes.

Figure 6.8 shows one of the filled in nodes expanded to reveal its sub-concepts. Each sub-concept inherits the attributes of the "parent" concept but has additional attributes potentially containing more valuable information about the crisis. In the example, there is a group of tweets that mention a shooting attack and two groups mentioning a fire or wildfire. There are also groups of tweets containing negative sentiment.

The analyst/decision maker also has the ability to trace back tweets of interest to the original source. The crisis concepts, instead of displaying the number of tweets, can display the source URLs. The analyst can select a URL to link back to the original text.

Thus, the application of FCA to aggregate source information into "crisis concepts" facilitates the decision maker by reducing information overload and focusing on crisis information along with its location. In ATHENA, the usability and interpretation are further improved by displaying crisis concepts in a map-based interface with additional filtering and search facilities.

6.6   Filtering and Searching

This section discusses the considerations and approaches to providing command and control operators with the necessary data they need in order to carry out their roles in a crisis situation when using the CCCID.

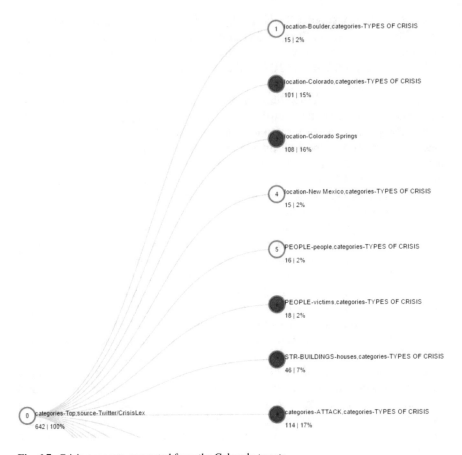

Fig. 6.7 Crisis concepts computed from the Colorado tweets

Fig. 6.8 Sub-concepts showing additional crisis information

There are various design considerations that should be made when providing filtering and searching capability and what could be arbitrarily applied depending on the role of the individual user or need. Two distinct phases of the crisis may also be recognized, which are during the time of the crisis, where information is more important in a relative context to the current situation, and the postcrisis phase where in-depth analysis and visualization has its importance.

6.6.1   Design Considerations

With a system such as the ATHENA CCCID being widely deployed, at any level, there is likely to be a large amount of data generated at any given time. This is simply due to the link between scale and distribution of crises, i.e., small crises occur often while large crises occur infrequently. As a result, the data flows in ATHENA could be reasonably steady but peak at certain times. The ATHENA consortium has found the following considerations to be of importance, particularly in the scope of crisis management.

6.6.1.1   Geographical Location

Clearly, it is important to ensure that the geographical location of any report coming in from the general public is available and as accurate as possible. Furthermore it is not necessarily realistic to envision only having one ATHENA system to manage all crises. ATHENA may be deployed on different scales. Where ATHENA is deployed, for instance, on a national scale, it would be feasible to have various groups or teams of operators monitoring and responding to particular geographical regions, just as call handlers and first responders would in any emergency response. With appropriate arbitrary filtering on geographical boundaries in place, a national deployment would result in similar behavior to a regional or even international deployment.

In such a scenario, it would be important for operators to be effectively registered or operational for only a subset of the reports and response in ATHENA—while there may be more strategic overviews at a higher level. Consideration should also be made in that there may be local and hyper-local responders using the mobile app or even the ATHENA CCCID (voluntary organizations, neighborhood watch, private sector security) who may require a geographically restricted view of activity.

6.6.1.2   Validity, Credibility, and Priority

It is sometimes the case that misinformation, or even disinformation, may be propagated throughout a system such as ATHENA for various reasons (hoax, malicious, political, errors) which is why the validity and credibility of incoming reports should be assessed. This may be a manual process, based on experience and expertise, though it could be automated in various ways.

In ATHENA, the focus with regard to validity and credibility is primarily on the source of the information, or user tier, which ranges from a full access operator level user, right down to an unregistered member of the general public. Both validity and credibility assessment or recommendations can be achieved based on whether other information has already been validated or rejected by the CCCID operators. The benefit of this comparative approach is the reduced risk of invading individual privacy.

Priority is slightly different in that it enables the system or operators to carry out validity and credibility assessment in the most efficient manner possible. The use of a triage approach to priority would focus the manual workload of the operators only on the most important tasks first.

Due to the risk of misinformation and disinformation enter the system, avoiding such abuse could be achieved through legislation supporting for individual registration for use or the automatic sharing of identity information by telecommunications companies when interacting with the system.

6.6.1.3   Information Source

Identifying the source of incoming information could be employed to aid the operators and automation processes with validation, credibility, and priority. An example of this may be an individual interacting directly with the ATHENA mobile app which represents a clear decision to provide information, whereas information coming in from social media is more likely to be misinterpreted. This classification should be available to the operator.

6.6.1.4   Ability to Perform Free-Text Searches

Any incoming bodies of unstructured text should be indexed appropriately to enable free-text search of the system. This should also work in combination with all other filtering options, giving the operator the power to find information within information. The difference between free-text search and all other types of filtering is that there is often no reason as to why it would need be arbitrarily applied and may be made available to everyone (conversely, an operator whose role is validation may need to see only reports that have not been validated yet and thus requires a very specific type of filter).

6.6.1.5   Information Aggregation

Due to the potentially steady, though sizable streams of information entering a system like ATHENA from uncontrolled sources, it makes sense (and may even be vital in many situations) that information is properly aggregated. ATHENA does this on-the-fly using a formal concept analysis algorithm. Though it does not stop at the actual aggregation process as operators need to be able to choose between accessing individual reports or aggregated reports to ensure information is not missed or even to validate the aggregation process. The reduction in information flow here can be highly beneficial; for example, there may be 15 reports representing the same instance with slight variations, showing the operator that one report instead of 15 will greatly improve their efficiency.

6.6.2   Requirements for Filtering and Searching During a Crisis

As a crisis unfolds, operators are unlikely to need functionality that provides them with full exploratory access to the ATHENA data set. Instead data needs to be geographically and temporally relevant to the individual operator and delivered in a way that enables them to carry out necessary responsibilities. Considering the types of filtering and searching capabilities discussed above, some are more important for operators during a crisis.

First of all is location. If an operator does not have any information regarding the location of those reporting the crisis, it makes their job extremely difficult or even impossible. This can be partly mitigated through aggregation as reports with and without locations may be aggregated together, providing a location for the set of aggregated reports; for example, a report regarding a vehicle crash in location A could potentially be aggregated with other reports at a similar time regarding a car fire and thus implying the fire is also at location A.

Such aggregation, or even association, between multiple incoming reports with various information, is difficult without considering the time information came in. During a crisis, it is more important for operators to see time in a relative way, such as '4 min ago.' Using time in this way and allowing operators to find data based on the last number of seconds, minutes, or hours can increase the speed in which they can filter through information.

The three measures of validity, credibility, and priority have similarities, but also serve individual purposes and complement one another. These classifications can be used by operators to change their perspective of the incoming information, i.e., see the current certain and uncertain state of the situation.

Finally, it is important, due to the potential for information overload, to provide operators with means of managing large volumes of information. As ATHENA employs aggregation processes to identify, combine, and corroborate incoming information, it enables operators to see and react to a more generalized picture of the unfolding crisis.

6.6.3   Requirements for Filtering and Searching Postcrisis

The needs and requirements for accessing data in the postcrisis phase are distinct from those required during the crisis. Postcrisis requires a view that is both highly generalized while also necessitating greater detail and interrogation capabilities. It is at this stage where data visualization is likely to provide most value. That said, location and time are still key players.

The element that is most vital with regard to having the ability to search at this stage is with crisis media (e.g., photos, videos, audio, etc.). Having the ability to quickly and easily visualize and traverse all media captured during the crisis is important for gathering a picture of the entire crisis and its response. This can easily be displayed on a timeline, as could the entire incident or media could be restricted to particular geographical regions.

Timeline visualizations can provide an insight into both when media was captured, likely revealing when key events occurred. They can also provide access to the pace at which new information came into the system at key stages during the crisis (i.e., in an earthquake situation these key events could be initial incident and aftershocks as well as building collapses) to provide an overview of the crisis.

Auditing all data, updates, and activities in the ATHENA system, whether internally or externally invoked, is extremely important for post-event analysis of the crisis as well as the response to the crisis. Every individual piece of data that enters the system should be recorded in a write-only audit. Every modification to the data or its state should also be recorded. The end goal here is ensuring that no piece of data goes missing, i.e., removing data from the CCCID view may be possible, but the audit will not miss anything. Visualizing and analyzing such data can provide valuable insights and lessons into the crisis but also the use of the mobile app and CCCID.

6.7   Results and Evaluation

Overall we can report some success in the processes we have developed when applied to the social media data from a real disaster. Our results also point to a number of promising lines for future development.

The main objective for the automated aggregation of information sources was to reduce information overload. In the example using the Colorado Tweets, the original 1200 tweets were firstly reduced to 642 by only considering those that contained a location and a reference to a crisis category. The FCA aggregation reduced this to 76 groups of tweets, each group containing tweets with the same location and crisis category. This represents a significant reduction in information points (88%) and, in this case, a final number of information points that would be manageable by an end user. However, if the starting point was a far greater number of information sources, additional steps may have to be taken to counter information overload, such as setting a minimum number of information sources per crisis concept. Better resolution of location names would also increase the level of aggregation and improve the quality of the crisis concepts. In the Colorado Tweets, there were nine 'versions' of Colorado Springs: Help Colorado Springs, Fame Colorado Springs, Colorado Springs, COLORADO SPRINGS, Davis Colorado Springs, North-western Colorado Springs, Technician Colorado Springs, The Colorado Springs and Hey Colorado Springs. Further work is required to improve the location recognition and extraction software to resolve identical locations and exclude erroneous locations.

The Colorado Tweets example also illustrated the ability of a crisis taxonomy to provide a useful 'drill-down' feature for the analyst. Figure 6.8 shows crisis concepts with increasingly specialized crisis categories, from 'Types of Crisis' to 'Natural Disaster' to 'ND-Wildfire.' The higher levels of aggregation are achieved at the more general levels in the taxonomy, but the analyst is able to then 'drill down' into ever more specialized sub-concepts that have fewer information sources but potentially more specific information about a specific crisis event. The most specialized concept may contain only a few information sources and at this point the analyst is able to trace back to the original sources to examine their text.

The sentiment analysis component identified sentiment correctly in more than 87% of the tweet corpus, while the remaining 13% showed how the analysis would benefit from further sentiment specialization. For example, in some instances, the sentiment analysis component identified negative sentiments expressed towards citizens. Upon manual inspection, these tweets were actually expressing emotions such as sadness, anger, and frustration towards the crisis event—this shows how further insight could be gained by extending the sentiment model to incorporate specialized sentiments such as anger and frustration, in the analysis, rather than only using the positive/negative scale (see Chap. 13).

6.8   Conclusion

Crowdsourcing and other online participatory practices are becoming increasingly important to emergency personnel. The benefits of harnessing social media data and the faster, localized information from technology enabling mass participation are significant. However, the risks and challenges of using such large pools of dynamic, unregulated material cannot be ignored. Here, we have reviewed the benefits and potential dangers of exploiting this information and the techniques that can be employed to combat them. We have discussed various techniques such as sentiment analysis, credibility assessment and priority assessment, and aggregation which aim to enhance the usefulness and reliability of the data and contribute to emergency assessment and response.

In this chapter, we have given an overview of the processes and systems used in ATHENA as a means of obtaining, analyzing, filtering, and presenting social media and crowd-sourced information. The result of this work is a set of powerful new capabilities which can be used in multiple ways to support the overall goals of the ATHENA project.

References

1. Dubey, R., Luo, Z., Xu, M., & Wamba, S. F. (2015). Developing an integration framework for crowd-sourcing and internet of things with applications for disaster response. In *2015 IEEE International Conference on Data Science and Data Intensive Systems* (pp. 520–524).
2. Sievers, J. A. (2015). Embracing crowdsourcing: A strategy for state and local governments approaching 'whole community' emergency planning. *State and Local Government Review, 47*(1), 57–67.
3. Alexander, D. E. (2014). Social media in disaster risk reduction and crisis management. *Science and Engineering Ethics, 20*(3), 717–733.

4. Oh, O., Agrawal, M., & Rao, H. R. (2013). Community intelligence and social media services: A rumor theoretic analysis of tweets during social crises. *MIS Quarterly, 37*(2), 407–426.
5. Gao, H., Barbier, G., & Goolsby, R. (2011). Harnessing the crowdsourcing power of social media for disaster relief. *IEEE Intelligent Systems, 26*(3), 10–14.
6. Narvaez, R. W. (2012). *Crowdsourcing for disaster preparedness: Realities and opportunities.* Master's thesis, Graduate Institute of International and Development Studies, Geneva, Switzerland.
7. Olteanu, A., Vieweg, S., & Castillo, C. (2015). What to expect when the unexpected happens: Social media communications across crises. In *Proceedings of the 18th ACM Conference on Computer Supported Cooperative Work & Social Computing* (pp. 994–1009). CSCW 2015, Vancouver, BC, Canada, March 14–18, 2015.
8. Andrews, S., & Orphanides, C. (2012). Knowledge discovery through creating formal contexts. *International Journal of Space-Based and Situated Computing, 2*(2), 123–138.
9. Andrews, S., & Orphanides, C. (2013). Discovering knowledge in data using formal concept analysis. *International Journal of Distributed Systems and Technologies, 4*(2), 31–50.
10. Domdouzis, K., Andrews, S., Gibson, H., Akhgar, B., & Hirsch, L. (2014). Service-oriented design of a command and control intelligence dashboard for crisis management. In *Proceedings of the 7th IEEE/ACM International Conference on Utility and Cloud Computing* (pp. 702–707). UCC 2014, London, UK, December 8–11, 2014.
11. de Assis, L. F. G., Herfort, B., Steiger, E., Horita, F. E. A., & de Albuquerque, J. P. (2015). Geographical prioritization of social network messages in near real-time using sensor data streams: an application to floods. In *XVI Brazilian Symposium on GeoInformatics* (pp. 26–37), Campos do Jordao, Sao Paulo, Brazil, November 29–December 2, 2015.
12. de Albuquerque, J. P., Herfort, B., Brenning, A., & Zipf, A. (2015). A geographic approach for combining social media and authoritative data towards identifying useful information for disaster management. *International Journal of Geographical Information Science, 29*(4), 667–689.
13. Cresci, S., Tesconi, M., Cimino, A., & Dell Orletta, F. (2015). A linguistically-driven approach to cross-event damage assessment of natural disasters from social media messages. In *Proceedings of the 24th international conference companion on World Wide Web.* ACM.
14. Verma, S., Vieweg, S., Corvey, W. J., Palen, L., Martin, J. H., Palmer, M., et al. (2011). Natural language processing to the rescue? extracting 'situational awareness' tweets during mass emergency. In *Proceedings of the 5th International Conference on Weblogs and Social Media*, Barcelona, Spain, July 17–21, 2011.
15. Yin, J., Lampert, A., Cameron, M. A., Robinson, B., & Power, R. (2012). Using social media to enhance emergency situation awareness. *IEEE Intelligent Systems, 27*(6), 52–59.
16. Castillo, C., Mendoza, M., & Poblete, B. (2011). Information credibility on Twitter. In *Proceedings of the 20th International Conference on World Wide Web* (pp. 675–684), Hyderabad, India, March 28–April 1, 2011.
17. Gupta, A., Kumaraguru, P., Castillo, C., & Meier, P. (2014). *TweetCred: A real-time web-based system for assessing credibility of content on Twitter.* CoRR abs/1405.5490.
18. O'Donovan, J., Kang, B., Meyer, G., Hollerer, T., & Adali, S. (2012). Credibility in context: An analysis of feature distributions in Twitter. In *2012 International Conference on Privacy, Security, Risk and Trust, PASSAT 2012, and 2012 International Conference on Social Computing* (pp. 293–301). Social Com 2012, Amsterdam, Netherlands. September 3–5, 2012.
19. Imran, M., Castillo, C., Diaz, F., & Vieweg, S. (2015). Processing social media messages in mass emergency: A survey. *ACM Computing Survey, 47*(4), 67.
20. Witten, I. H., Frank, E., & Hall, M. A. (2011). *Data mining: Practical machine learning tools and techniques* (3rd ed.). San Francisco, CA: Morgan Kaufmann.
21. Ganter, B., & Wille, R. (1999). *Formal concept analysis—Mathematical foundations.* New York: Springer.

Chapter 7
The ATHENA Mobile Application

Chi Bahk, Lucas Baptista, Carly Winokur, Robin Colodzin, and Konstantinos Domdouzis

7.1 Introduction

For large crisis events, mobile devices and applications can contribute significantly in data exchange amongst citizens, first responders, police, and Law-Enforcement Agencies (LEAs). Examples of geolocation-enabled mobile crowdsourcing applications, such as Love Clean Streets [1] and HealthMap's Outbreaks Near Me [2], reveal the strength of combining the social web ('Web 2.0') and smartphones to provide levels of engagement and participation to their local and wider communities [3].

Hagar [4] has developed the term 'Crisis Informatics' that refers to socially and behaviourally conscious use of Information and Communication Technologies (ICT) during a crisis. The term was later extended by Hughes et al. [5] and Palen et al. [6]. During a crisis, the exchange of information is extremely critical and various ICT can facilitate this exchange [7]. Crisis events, such as the September 11 attack and Hurricane Katrina, underlined the significance of information provision from citizens to the authorities and showed that there must not be reliance on a single communication infrastructure [8].

A mobile application that maps threats and helps people in an evacuation process is SmartRescue. The project utilises embedded sensors in smartphones, such as audio-visual sensors, but also pressure, temperature, light, accelerometer sensors, and GPS. The SmartRescue platform links crisis responders to users who are close to the hazard, for responders to communicate the location of the hazard, receive threat

C. Bahk (✉) • L. Baptista • C.Winokur • R. Colodzin
Epidemico, Boston, MA, USA
e-mail: chi@epidemico.com

K. Domdouzis
CENTRIC, Sheffield Hallam University, Sheffield, USA

© Springer International Publishing AG 2017
B. Akhgar et al. (eds.), *Application of Social Media in Crisis Management*,
Transactions on Computational Science and Computational Intelligence,
DOI 10.1007/978-3-319-52419-1_7

pictures in real time through a map, and develop evacuation plans based on the evolving threat. The platform is a decision support tool for those who are affected by a disaster or are responsible for its handling. The platform includes a number of information senders and receivers. In this case, users can send sensor information through the use of mobile phones or subscribe to receive sensor information from others [9].

Ushahidi is a crowdsourcing software that was produced after the post-election violence in Kenya in December 2007 [10]. Ushahidi allows users to report an event through its website, SMS text messaging, email, and through a hotline. The reports are aggregated on an interactive map with which users can be informed of reports around them. Two years after the event in Kenya, Ushahidi was also used during the Haiti earthquake. Specifically, volunteers at Fletcher School of Tufts University used the mapping software to develop a live map of Haiti just hours after the strike of the earthquake. Days later, a coalition of partners created a short code for Haiti—a project called Mission 4636—to allow anyone in the country to text in their location and most urgent needs. Thousands of volunteers rallied to translate these messages from Haitian Creole to English. Volunteers at Fletcher School would then identify the most significant translated messages and add them to the live map of Haiti [11]. Ushahidi has since been used for reporting human rights violation, disease surveillance, and disaster response [12].

Gao et al. [13] underline the significance of developing coordination protocols and mechanisms to manage governmental and non-governmental organisation activities. Research has shown that it is possible to develop community crisis maps based on information received from social media and introduce an interagency map, so that organisations can exchange information. Social media has played a crucial role during natural disasters as a means of data propagation that can be used for disaster relief. After the catastrophic Haiti earthquake in January 2010, the public posted a large number of text, photos, and videos about their experiences during the earthquake to social media sites, such as Twitter, Facebook, blogs, and YouTube. The Red Cross received eight million dollars directly via text donations [13].

Yuan and Detlor [14] present an intelligent mobile system used for crisis response. An intelligent Crisis Response System (CRS) should be characterised by six sequential steps: Monitoring & Reporting, Identification, Notification, Organization, Operation, and Assessment & Investigation. The realisation of these steps is based on the use of a distributed topology of advanced information technologies that uses a network of mobile devices that remain active, monitor the environment, and exchange data with users and with other systems. An example of the use of such a crisis response system is the detection of hazardous substances on trains. Sensors placed on a train detect fire and notify the train's engineer. Local police is notified over mobile communication devices. Software automatically searches the rail company's database looking for the train's cargo information. The jurisdictional Hazardous Materials Coordinator is then contacted. Local residents are notified about possible evacuation plans through radio, TV, and automatic dialling. Qualified personnel such as firefighters, medical personnel, police force, and volunteers are contacted over mobile communications in order to take on specific roles. A central command centre forwards information to the right parties. This information includes guidelines on evacuation. Crisis-related information is collected from mobile channels and all this information is classified with appropriate keywords. Statistics are generated about the damage, life loss, and injury [14].

A very significant reason for lack of coordination between the police, LEAs, and the Search and Rescue Teams is the lack of interoperability between communications equipment. The 9–11 commission report showed that incompatible technology and the uncoordinated use of frequency bands were the main reasons for poor communication during emergency response and recovery operations. Another problem is the strong dependence on terrestrial communications, such as landline and cellular telephony, as well as infrastructure-based Land Mobile Radio (LMR) networks [15]. Such findings indicate that different technologies must be available for usage during a crisis. One such example is the Wireless Mesh Network (WMN). This type of network has the ability to self-organise and self-configure. The nodes of a WMN have the ability to automatically detect network nodes and maintain connectivity in an ad hoc fashion by using ad hoc routing protocols [16]. Wireless mesh networks also have the ability to adjust to changing conditions and thus, they are characterised by high-level fault tolerance and robustness [17].

It is also important to underline the significance of social media in the management of crises. During large crises, the use of social media increases. A survey by the American Red Cross revealed that 69% of adults believe that social media sites should be monitored by emergency responders and 74% expect search and rescue services to answer social media calls for help within an hour [18]. Social media are very useful after crises as they enable citizens to share crisis information and even provide emotional support [19,20]. It can be concluded that a mobile communication system is required that will combine the flexibility of using different communication technologies, constant availability, and the use of social media for communicating crisis-related information. The design of the ATHENA mobile application was based on the consideration of these factors and this design is presented in more detail in the next sections.

The development and testing of the user-friendly ATHENA mobile application (henceforth referred to as 'the ATHENA App') responds to increased recognition that crucial aspects of communication between the first responders, police, and LEAs with civilians during large crisis situations require enabling smartphone users to request help, report observations, and receive critical information. In this chapter, we take a close look at the main features and functionalities of the app itself, and at the ways in which it interfaces with the Command, Control, Centre and Intelligence Dashboard (CCCID). This allows citizens to send user reports and receive information from the relevant central crisis command unit.

7.2 Description of the ATHENA Mobile App and User Types

The design of the ATHENA App was predicated on a number of known requirements. Uppermost amongst these was the app supporting two types of users: the trusted user and the citizen user. There are two tiers of trusted users (Tier 1—top tier, Tier 2—lower tier). Tier 1 users include first-responders and bronze (Operational), silver (Tactical) and gold (Strategic) command, while Tier 2 includes such people as utilities controllers or members of local resilience forums. Trusted users are required to authenticate themselves through a pin-code login. Citizen users do not need to be authenticated.

All users can add Emergency Information via Settings, such as name, blood group, and emergency contact phone number Emergency Information can be automatically attached to outputs, such as Reports or Help messages.

Users can add 'My Important Locations' (such as home, work, locations of relatives, schools of children) via the app's Settings. Such locations will only appear on the user's Crisis Map. A user will receive a notification when a validated ATHENA Report is posted near their location or one of their important locations. When a user first opens the app, he/she is shown a user-friendly summary of the privacy statement. This statement is displayed in an interactive and visual way, accompanied with icons and short, easy-to-understand statements. Once the user agrees to the privacy state summary, he/she is taken to the Home Screen with all main features of the app: Map View, List View, Send a Report, and I Need Help. Each button is accompanied by a clear and intuitive symbol. Users are able to access Settings via a menu located on the left side. Under settings it's possible to change the app features including the Map Provider, Map radius, Language, amongst other data visualisation preferences.

When the user accesses the application through Android or iOS, he/she is taken to the Home Screen, which is shown in Fig. 7.1. The Android and iOS versions of the ATHENA app have support for user accounts and profile information. A simplified version of the application is available for other mobile operating systems (e.g. Windows or Blackberry).

The user can open the Side Menu which includes all the main features of the app, such as the Settings option, the Icon & Colour Chart, and the link to the ATHENA Social Media pages. This facility is shown in Fig. 7.2.

A Shortcut menu is provided for the two urgent features—Send a Report and I Need Help. On the non-urgent features Map and News, there is a red! icon in the bottom centre of the screen. When it is clicked, the Shortcut Menu appears, at which point it is possible to select between the two urgent features.

The User Settings option, which can be accessed from the Side Menu, allows the user to:

- Add My Important Places (optional, login required)
- Add My Emergency Information, such as name, blood type, and a government ID number (optional, login required)
- Emergency contact (optional, login required) (Fig. 7.3)
- View My Reports
- Set preferences

  - Opt-out of sharing one's location
  - Language settings
  - Choose Map service (Google or Open Street Maps)
  - List view option (Preview or Full)
  - Set default view radius, to filter reports by distance from current location

- Log in or create account

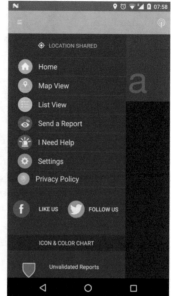

Fig. 7.1   Home screen—iOS and Android

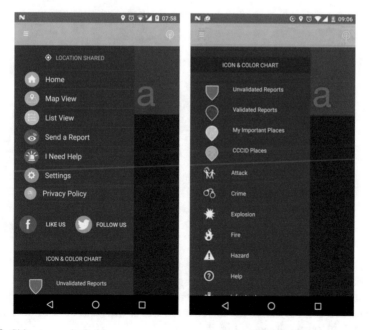

Fig. 7.2   Side menu—Android

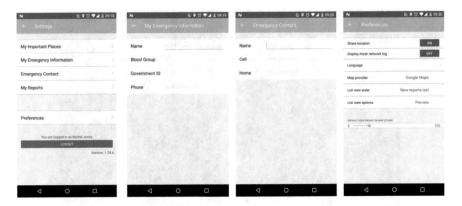

Fig. 7.3 Features under Settings

Fig. 7.4 User login in settings

As stated above, the ATHENA App supports three types of users: Tier 1, Tier 2, and Public (Citizens). The user's accredited type will determine his/her clearance level and the content that is made visible to them. This can be done through the Login in Settings, as shown in Fig. 7.4.

7.3   ATHENA Mobile App Features

The ATHENA App is characterised by a number of features with different function-alities. These features are the Mobile Crisis Map (or the "Map View"), the List View, the 'Submit a Report' feature, the 'I Need Help' feature, and the Messages (for trusted users only) feature. Each of these functionalities is described below.

7.3.1   ATHENA Mobile Crisis Map

The Mobile Crisis Map is implemented through the 'Map' feature incorporated in the ATHENA App. Users can access this feature from the Home Screen, the Side Menu, or from the bottom left Map icon in the News View. Whenever any particular user clicks on this menu item, they are immediately taken to a map centred at their current location with a 20 km radius view.

7.3.1.1   Map Pins

The Mobile Crisis Map uses a number of different types of pins. These are dis-played in Table 7.1.

For trusted users, there is differentiation of the validation and read statuses as shown in Table 7.2.

When any pin appearing on the map is clicked, a text box is displayed indicating the title of the report, and the date/time that it was posted. When this text box is clicked, detailed content of the incident appears. The Mobile App Map only dis-plays one summary report of an aggregated set (as determined by CCCID) for rea-sons of usability (Fig. 7.5).

Table 7.1 Types of pins

Type of pin	Pin used
1. User or CCCID reports	This pin is displayed together with crisis symbology
2. My important locations	
3. CCCID important locations	

Table 7.2 Types of pins based on Report Status

	Type of pin	Pin used
Validation status	1. Validated	
	2. Unvalidated	
	3. Rejected	
Read status	1. Read	
	2. Unread	

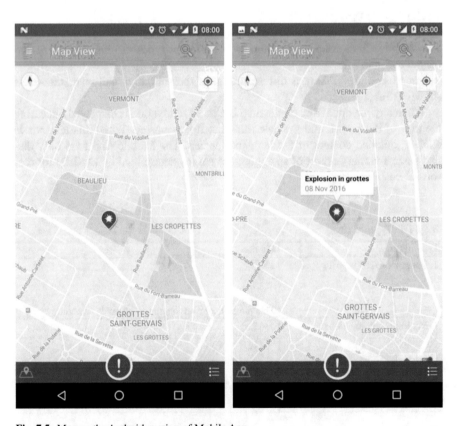

Fig. 7.5 Map on the Android version of Mobile App

7.3.1.2 Searches and Filters

A number of filters can be applied in the Mobile App so that pins that match certain criteria, such as category of crisis incident (e.g. fire, attack, crime, etc.). Other filtering criteria refer both to the type of information to be displayed (Reports, CCCID Locations, CCCID Reports) and the media by which it is represented (Text, Image, Audio, Video). There are additional filters for trusted users, such as Read Status, Validation Status, Credibility Status, Priority, and Clearance Level.

7.3.1.3 Map Services: Google Map and Open Street Maps

For reasons of accuracy, the application uses two mapping services (i.e. Google Maps and Open Street Maps) (Fig. 7.6).

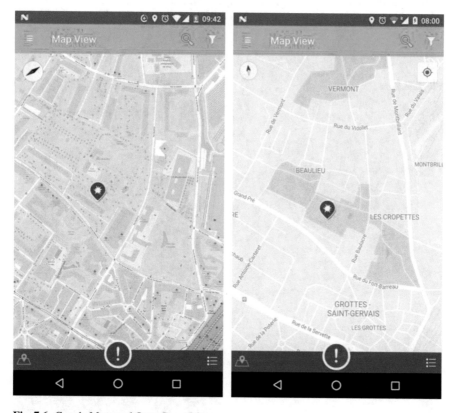

Fig. 7.6 Google Maps and Open Street Maps

7.3.1.4 Zones

CCCID operators can highlight geographic areas of significance. These areas are displayed on the Mobile Crisis Map as Zones. By clicking on the displayed zone, the user can see its description. The CCCID operators can send push notifications to all the users whose last registered positions are within a zone (Fig. 7.7).

7.3.2 List View

In the List View, users can view all the Reports for which they have been given clearance, according to user tiers. The List View can be shown in two different view settings: either a Full List View (which includes a number of advanced characteristics, such as all text and media) or as a Preview List View (which contains fewer characteristics, displaying only a thumbnail indicating the type of media attached, title, and a short version of the description provided).

Users can filter Reports. The resulting list reflects the criteria in the filter settings. When a user clicks into a single Report, the Incident Details screen provides an interactive map that can be clicked, taking the user to the Map view of the specific

Fig. 7.7 Zones

Report. Reports can be pinned to the top of the list by CCCID operators or aggregated to only have one summary Report visible to the users (Fig. 7.8).

A Headline that is posted by the CCCID is highlighted at the top of the List View. If there are multiple headlines, the user can scroll through them in turn. Examples of headlines in the ATHENA App are shown in Fig. 7.9.

7.3.3   'Send a Report' Feature

When a user clicks 'Send a Report', they are led to an appropriate Report screen where they can select the Category of the Report, edit the location, date and time, input text, and attach media (picture, video, audio). All submitted reports are sent to CCCID for review and validation. The user must acknowledge he/she is sending any saved user information to the CCCID operator along with the report. When a report is submitted, the user receives a confirmatory email (Fig. 7.10).

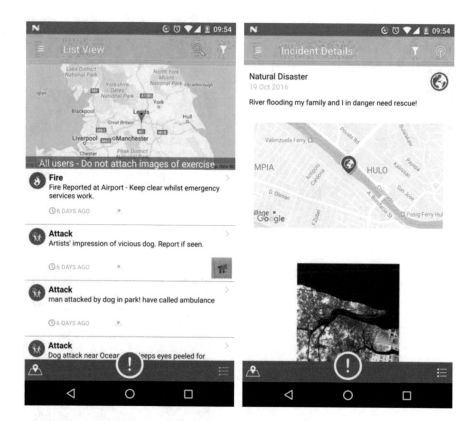

Fig. 7.8 News list and incident details

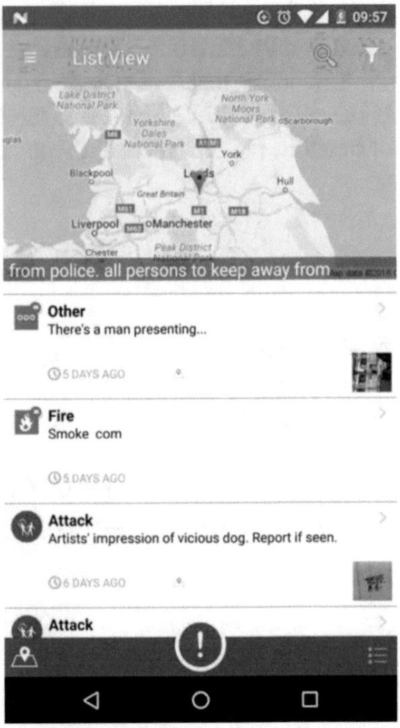

Fig. 7.9 Headlines

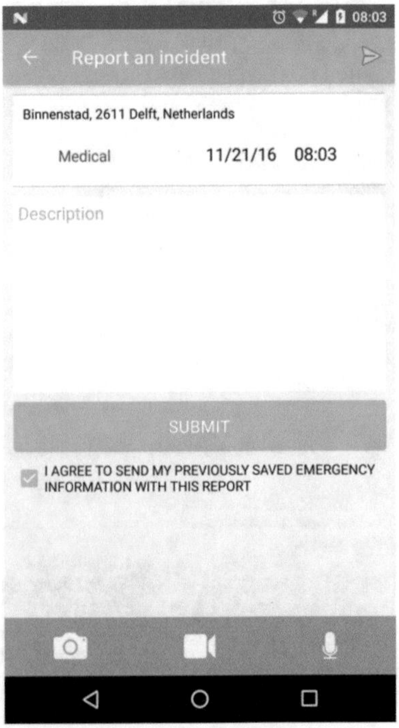

Fig. 7.10 Report form

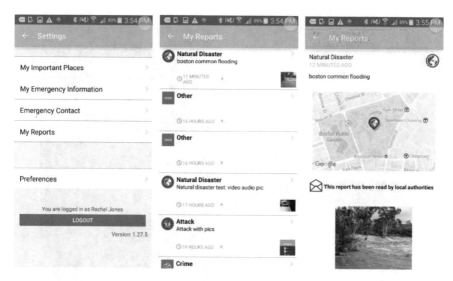

Fig. 7.11   My Reports

A user is able to view My Reports in Settings. In this case, a user can see his/her submitted reports and a notification whether or not it has been read by CCCID operators (Fig. 7.11).

7.3.4   'I Need Help' Feature

The 'I Need Help' feature consists of the Life Support system of the ATHENA App. This system allows for emergency messaging during a crisis in instances, including when traditional forms of communication are unavailable.

The 'I Need Help' feature offers specific options to the user. Firstly, it allows the user to call the local emergency number. The user can also send an emergency message to his/her emergency contacts with the user's current location. A Help Request can also be sent to CCCID.

The user can read the read status of their Help Requests in Settings, under My Reports (Fig. 7.12).

In cases where there is an absence of cellular connection, the user has the option of recording a text or voice message, which will be sent to CCCID through Internet connection. If there is problem with connecting to the Internet, the message is sent as soon as Internet or cellular connection is restored. When the request for help is finally sent, its read status can be checked by the user in Settings, under My Reports (Fig. 7.13).

If cellular and internet communication is lost or is non-existent, the Mobile App offers the feature of a Mesh Network—a system that pulls together a matrix of

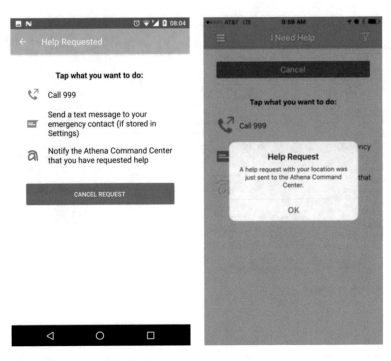

Fig. 7.12 Help Request—with cellular connection

Fig. 7.13 Help Request—
with no cellular connection

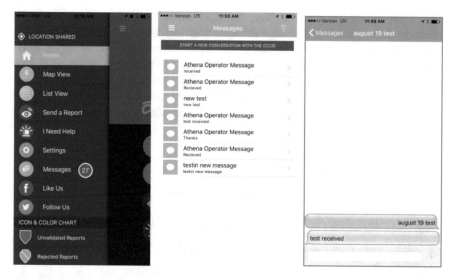

Fig. 7.14 Messages

mobile devices that have the Athena App installed. Each device can act as a router capable of passing on reports from one device to another. However, owing to the constant evolution of the technology, full functionality of the Mesh Network feature is not guaranteed. Thus, the user would need to actively search for other devices in order to transfer their request for help.

7.3.5   Messages (Trusted Users Only) Feature

The Message feature allows trusted users (Tier 1 or 2) to communicate directly with the CCCID operators. This feature is not available to Tier 3 users or non-logged in users. The application will notify the users when there are new, unread messages from the CCCID (Fig. 7.14).

7.4   Conclusions

The ATHENA Mobile App is an integral part of the ATHENA system for crisis management. The Mobile App offers a number of functionalities to its users. More specifically, it allows the users to access a map of crisis events with different search and filtering options and the depiction of dangerous and safe zones. The app provides a list view of crisis incidents and the option for users to report a crisis incident. Additionally, users can request immediate help through the 'I Need Help' feature, which allows users to make a phone call to a local emergency number, or send a message to the user's emergency contacts with the user's location. Finally, the 'Messages' feature allows trusted users to directly communicate with CCCID operators. The application's design was based on the capabilities of different mobile technologies and options that should be provided to citizens during a large crisis incident. A number of settings ensure the validity of information collected from citizens, even allowing a help request to be transmitted when Internet or cellular communications are unavailable.

References

1. URL1. (2016). *Love clean streets*. Retrieved July 10, 2016, from http://www.bbits.co.uk/ Home/LoveCleanStreets.
2. URL2. (2007). *Outbreaks near me*. Retrieved July 10, 2016, from http://www.healthmap.org/ outbreaksnearme/.
3. Kamel Boulos, M. N., Resch, B., Crowley, D. N., Breslin, J. G., Sohn, G., Burtner, R., et al. (2011). Crowdsourcing, citizen sensing and sensor web technologies for public and environmental health surveillance and crisis management: Trends, OGC standards and application examples. *International Journal of Health Geographics, 10*, 67. doi:10.1186/1476-072X-10-67.
4. Hagar, C. (2007). The information and social needs of farmers and use of ICT. In B. Nerlich & M. Doring (Eds.), *From Mayhem to meaning: Assessing the social and cultural impact of the 2001 foot and mouth outbreak in the UK*. Manchester, UK: Manchester University Press.
5. Hughes, A., Palen, L., Sutton, J., Liu, S., & Vieweg, S. (2008). "Site-seeing" in disaster: An examination of on-line social convergence. In *Proceedings of the 5th International Conference of the Information Systems for Crisis Response and Management (ISCRAM)*, Washington, DC, May 2008.
6. Palen, L., Vieweg, S., Liu, S., & Hughes, A. (2009). Crisis in a networked world: Features of computer mediated communication, in the April 16, 200 Virginia Tech Event. *Social Science Computer Review, Special Issue on e-Social Science, 27*(5), 1–14.
7. Kavanaugh, A., Sheetz, S. D., Quek, F., & Joon Kim, B. (2010). Cell phone use with social ties during crises: The case of the Virginia Tech Tragedy. In *7th International Conference on Information Systems for Crisis Response and Management: Defining Crisis Management 3.0*, Seattle, WA.
8. May, A. (2006). *First informers in the disaster zone: The lessons of Katrina*. Washington, DC: The Aspen Institute.
9. Radianti, J., Dugdale, J., Gonzalez, J. J., & Granmo, O. -C. (2014). Smartphone sensing platform for emergency management. In *Proceedings of the 11th International ISCRAM Conference–University Park*, Pennsylvania, USA, May 2014.
10. Kinyanjui, K. (2008). *Kenya: Citizens' reporting tool comes in handy*, [Blog]. Retrieved from http://allafrica.com/stories/200801150990.html.
11. Meier, P., & Munro, R. (2010). The unprecedented role of SMS in disaster response: Learning from Haiti. *SAIS Review, 30*(2), 91–103.
12. McClendon, S., & Robinson, A. C. (2012). Leveraging geospatially-oriented social media communications in disaster response. In *Proceedings of the 9th International ISCRAM Conference* (pp. 2–11), Vancouver, Canada.
13. Gao, H., Barbier, G., & Goolsby, R. (2011). Harnessing the crowdsourcing power of social media for disaster relief. *IEEE Intelligent Systems, 26*, 10–14.

14. Yuan, Y., & Detlor, B. (2005). Intelligent mobile crisis response systems. *Communications of the ACM, 48*(2), 95–98.
15. Portmann, M., & Pirzada, A. A. (2008). Wireless mesh networks for public safety and crisis management applications. *IEEE Internet Computing, 12*, 18–25.
16. Royer, E. M., & Toh, C. K. (1999). *A review of current routing protocols for ad hoc mobile wireless networks. IEEE Personal Communications Magazine, 6*, 46–55.
17. Portmann, M. (2006). Chapter 16. Wireless mesh networks for public safety and disaster recovery applications. In Y. Zhang, J. Luo, & H. Hu (Eds.), *Wireless mesh networking architectures, protocols and standards* (pp. 545–576). Boca Raton, FL: Auerbach Publications.
18. American Red Cross. (2010). *Web users increasingly rely on social media to seek help in a disaster*. Retrieved October 6, 2016, from http://www.prnewswire.com/news-releases/web-users-increasingly-rely-on-social-media-to-seek-help-in-a-disaster-100258889.html.19. Choi, Y., & Lin, Y. (2009). Consumer Responses to Mattel Product Recalls Posted on Online Bulletin Boards: Exploring Two Types of Emotion. *Journal of Public Relations Research*, 21(2), 198-20720. Stephens, K., & Malone, P. (2009). If the Organizations Won't Give Us Information…: The Use of Multiple New Media for Crisis Technical Translation and Dialogue. *Journal of Public Relations Research*, 21(2), 229-239.

Chapter 8
Standardization to Deal with Multilingual Information in Social Media During Large-Scale Crisis Situations Using Crisis Management Language

Kellyn Rein, Ravi Coote, Lukas Sikorski, and Ulrich Schade

8.1 Introduction

One of the most effective uses of social media in crisis situation is for the distribution by emergency services, government agencies, and others of reliable information about the nature of the crisis. The panic and confusion that often occurs in the early stages of a crisis can be minimized through such messaging. The value of this role of social media in such cases cannot be underestimated.

While some crises such as political unrest, wildfire, or outbreaks of highly infectious diseases have a buildup over time, many are unanticipated, triggered by a sudden, catastrophic event such as a terrorist attack, earthquake, or tsunami. In such catastrophic crises, there is a period of time at the beginning of a situation where emergency and rescue services are not yet on the scene. Here social media may provide a useful information flow in the opposite direction, namely, from witnesses on the scene to emergency services and law enforcement during the crucial early stages. In this age of smartphones and social media, the availability of instantaneous, on-site information from witnesses, victims, and others close to the scene about the unfolding crisis can make Twitter, Facebook, and others such social media applications sources of great potential for emergency response and crisis management.

However, the use of social media sources for decision-making in a crisis situation can be a double-edged sword. By its open nature, social media potentially offers insight on the ground at the very early stages of an emergency situation that would otherwise not be available (see Chaps. 2–4). However, this openness and accessibility may result in complications and potential pitfalls; these have to be

K. Rein (✉) • R. Coote • L. Sikorski • U. Schade
Fraunhofer FKIE, Wachtberg, Germany
e-mail: kellyn.rein@fkie.fraunhofer.de; ravi.coote@fkie.fraunhofer.de; lukas.sikorski@fkie.fraunhofer.de; ulrich.schade@fkie.fraunhofer.de

© Springer International Publishing AG 2017
B. Akhgar et al. (eds.), *Application of Social Media in Crisis Management*,
Transactions on Computational Science and Computational Intelligence,
DOI 10.1007/978-3-319-52419-1_8

identified, analyzed, and weighed carefully when making the decision to use information derived from social media sources.

One issue is that hundreds, if not thousands, of eyewitnesses generate a huge volume of information within minutes of the events. As soon as friends, family, and others receive these communications, they are forwarded (e.g., re-tweeted) to yet more contacts, propagating at a tremendous rate. Along with re-tweets of eyewitness information, there are tweets (and re-tweets) generated by noninvolved persons expressing horror, offering assistance, theorizing as to the cause, etc., which contain no useful information for emergency services.

The volume of information flowing through social media during a crisis necessitates, as we shall see, an automated solution for at least initial categorization of that information; one must always keep in mind that social media sources are not just activated in emergency situations, but are continuously operating with communications on all levels and on a wide range of topics that have nothing to do with the crisis at hand. As we will see later in this chapter, the Paris bombings generated *over four million tweets* worldwide within the first 24 h alone. This volume makes it nearly impossible for humans to evaluate which incoming messages contain valuable information and which do not. Although communications about the crisis will generally come to dominate social media shortly after the triggering event, there will further continue to be 'background noise' in the form of noncrisis communications which needs to be filtered out.

In addition to background noise, social media may be considered, on a certain level, to be a very prolific rumor mill (see Chap. 6 for information validation). A significant amount of Twitter communications, for example, consist of re-tweets, that is, the forwarding of messages created by other users. This has two serious effects for crisis management: volume and credibility of information. In the first case, there is a significant volume of repetitive messages that bring no new information but which require processing time to determine their repetitiveness; this can place a strain on any system which is processing the messages coming through. In the second case, retweeting occurs whether the information is credible or not, thus a significant number of social media messages may contain questionable and inaccurate information—even, as we shall see later in this chapter, intentionally malicious disinformation—all of which muddy the waters for crisis management teams.

When the bombs exploded in Brussels in March 2016, the world looked on in horror at the carnage wrought at Brussels International Airport and Maelbeek metro station in the city center. The very first reports of damage, death, and injury which appeared came from eyewitnesses via social media and included photos, films, and text messages. Clearly, in such circumstances, social media can be an incredible source of information for first responders, but that is only if they can make sense of the information received.

The Brussels attacks also highlight one issue very clearly: there are many languages with which people communicate. While some crises may be limited to geographical areas in which a single natural language is so dominant that nearly all of the pertinent communications about the crisis are in that single language. However, in some locations, such as Brussels, the likelihood of relevant social media communications being monolingual is nil, and not simply because Belgium has three official languages. Additionally, in large-scale crisis, such as the 2010 earthquake in Haiti or the 2004 Indian Ocean tsunami, there are often external aid organizations such as Red Cross and Medécins sans Frontières which help to deal with the aftermath. In Haiti there were groups from such diverse countries as China, Israel, Iceland, the USA, and Korea, to name but a few. Even if a common language such as English was used among these groups, it was not one of the two native languages in Haiti, thus guaranteeing at least trilingual social media communications. Therefore, development of strategies to deal with multiple languages in crisis situations is a no trivial consideration in many parts of the world.

Thus, it is important to find a way to automate a process to deal with multiple languages to locate important information generated in each of these languages, to filter out non-eyewitness information, and identify content which is relevant for responders.

In the following sections, we will discuss these topics at more length and offer some suggestions for solutions to the problems identified. Along the way, we propose a methdology which not only provides a solution to all of the issues mentioned above, but also allows for ease of communication between multinational responders in large-scale disaster situations.

8.2   Numbers of Messages and Speed of Dissemination

One of the most significant issues in the use of social media as a source of actionable information is crisis lies in the sheer volume of messages generated. In her article "The power of one wrong tweet" [1], CNN correspondent Heather Kelly described how hackers took over the Associate Press Twitter account, posting a tweet claiming that there had been a several explosions at the White House and that President Obama was injured. Although the tweet was taken down after only a few minutes, during that short time it was re-tweeted more than 3000 times. To put this re-tweet rate into perspective, if one assumes a human operator would need only 30 s to read and make an initial decision about the type and pertinence of each message, looking through this number of tweets would require 90,000 s, which is 1500 min or 25 h— far longer than the time needed to generate them. In a post on the Washington Post's Monkey Cage blog [2], Alexandra Siegel, a researcher at NYU's Social Media and Political Participation (SMaPP), describes her analysis of *more than 4 million tweets* which were sent in the 24 h following the Paris attacks in March 2016. Sifting through and verifying these would be a task of many degrees of magnitude higher than the previous example—at 30 s per tweet, 4 million would require nearly 4 years of 24/7 processing for completion. However, Siegel's analysis included tweets which generated outside of the immediate area of the tragedy and included messages of sympathy and support from around the globe.

A more realistic picture of the volume of tweets in a crisis is given by Tersptra et al. [3] who analyzed tweets generated during a storm which hit a festival at

Pukkelpop in Kiewit, Belgium in August, 2011. In total, they looked at nearly 97,000 tweets which appeared in a 12-h time span just prior to the storm and immediately following it, with the bulk (some 94,000) occurring in the time period from 6:15 to midnight. According to their data, the peak rate was 576 tweets a minute, about 3 h after the storm.

Although this last example is of a relatively small, relatively unknown nature (in comparison to the Paris or Brussels attacks, the Indian Ocean tsunami, and the earthquake in Haiti), the volume of messages produced just over Twitter gives a sense of the scale of processing needed to make sense from this information. It behooves us to examine if there may be ways to determine the most useful of the information to allow for more rapid processing to support decision-making.

8.3 Filtering by Sources and Types of Social Media Communications

Even if there were a way to efficiently process this volume of tweets in a timely manner, it doesn't necessarily make sense to do so, as not all of the communications contain actionable information. For example, a significant number of tweets are 're-tweets,' basically the republication of an original tweet by a Twitter. In other words, these are simply copies, which seldom contain new information, unless a comment has been added by the re-tweeter. Identifying and eliminating re-tweets would generally reduce the number of tweets to be analyzed: in the case of the above mentioned tweet about President Obama being injured, elimination of re-tweets would have reduced the number of tweets to be analyzed from over 3000 down to the single original message.

Similarly, in the previously mentioned study about the Pukkelpop storm [3], when Terpstra et al. examined the immediate period of the storm, they found that there were 674 tweets in the first hour which reported on damage which had occurred, but of those, 64% were re-tweets. They also noted that one single tweet was re-tweeted 119 times during their tracking period.

In addition to the large number of repetitive messages, there are differences in the sources of information and the relevance to the requirements of the management team in a given crisis. Otneau et al. [4] performed an analysis of Twitter communications in 26 separate crises, which were either 'human-induced' (accidental or intentional) or natural disasters (including floods, earthquakes, typhoons, etc.); from this analysis, they identified seven basic categories of tweets common to all crises:

1. Affected individuals, which covered information concerning 'deaths, injuries, missing, trapped, found or displaced people' as well as updates about the tweeters themselves or their families, etc.
2. Information on infrastructure and utilities such as conditions (damaged, restored, operational) of buildings, roads, utilities, etc.

3. Donations and volunteering, such as requests or needs, queries, offers of supplies, shelter, or volunteer services
4. Caution and advice, including warnings which were issued or lifted, tips, etc.
5. Statements of sympathy and emotional support
6. Useful information not covered above, such as references to websites or telephone numbers for specific support activities
7. Not applicable or related to the crisis, including not readable

Of these seven categories, only the first two will be of major interest to crisis managers for information about injuries and damage that might affect operations in the critical early stages of a catastrophic crisis. For example, information that a bridge has collapsed will result in re-routing of emergency vehicles, knowledge of the locations, and messages about the number of injured citizens will assist in prioritizing and assigning medical assistance. The third category (donations and volunteering) may be of interest at later stages. Categories 4 and 6 are less likely to come from individuals at the scene, but rather from government and nongovernment sources, as well as media sources or businesses.

Breaking the complete corpus down into the relative frequencies of each category provided the following results: the largest percentage (32%) of tweets were of the 'useful information' (category 6); messages of both category 1 'affected individuals' and category 5 'sympathy and emotional support' each appeared roughly 20% of the time, with 'donations and volunteering' (category 3) as well as 'caution and advice' (category 4) coming in at 10% each, 'infrastructure and utilities' (category 2) at 7%.

Thus, the two most pertinent categories for early crisis management ('affected persons' and 'infrastructure and utilities') together comprised only 27% of the total messaging across crises.

(It should be noted that the breakdown varied across types of crises, depending, for example, on the type of crisis, including whether it was 'progressive,' i.e., slow buildup or 'instantaneous')

Olteanu et al. [4] additionally categorized the types of sources from which the tweets originated: eyewitnesses, government, nongovernmental organization, business (defined as "for-profit" but excluding news organizations), traditional and/or internet media (television, radio, news organizations, blogs, etc.), and outsiders (individuals who are not personally involved or affected by the event).

Messages from eyewitnesses accounted for an average of 9% of the communications, while traditional and Internet media accounted for the largest cluster at an average of 42% of the tweets. Unsurprisingly, communications from government and NGO sources together counted for an average of just 5%; this is to be expected because the communications released by both of these sources would have been verified for accuracy, as well as focused on specific types of information for the general public.

The second largest category of tweets at an average of 38% of the total originated from 'outsiders,' composed of individuals neither directly involved in nor affected by the crisis; these tweets tended to be of the category 'statements of sympathy and emotional support.'

The conclusion of the researchers is that a significant number of tweets may be discarded as not useful or uninteresting content for the management of the crisis:

> ...For instance, if an analyst or an application focuses on content that is not present in mainstream or other Internet media sources, and wants to exclude content provided by outsiders who are not affected by the crisis, then it will have to skip through 80% of the tweets on average. (ibid., p. 1005)

When filtering by source types, one may also significantly reduce the number of tweets to be processed: for example, if only eyewitness information is being sought, then on average over 90% of tweets may be ignored simply by identifying whether the information in the tweet is first person. This dramatically reduces the processing load.

Finally, we return to the issue of redundant or duplicate messages (re-tweets) with which we started this section. The study by Ortneau et al. mentions such messages only in passing in a single paragraph under the heading 'content redundancy.' Their definition of redundancy relies on identifying commonalities between messages, which were considered to be 'near-duplicates, if their longest common subsequence was 75% or more of the length of the shortest tweet.' One can assume that this 75% rule would allow for the insertion by the re-tweeter of a comment or perhaps an additional hashtag. Details were not provided for all source types and information categories, but there is some insight into how redundant information can dominate. For example, the top 3 messages and their near-duplicates from government and nongovernment agencies accounted for over 20% of tweets from these sources, whereas in the information category 'infrastructure and utilities' the top three messages accounted for roughly 12–14% of the total. This clearly signals that identifying and eliminating redundant messages may be a strategy to reducing overall processing time and resources in order to provide crisis managers and decision-makers with actionable information.

One thing is thus very clear: that in order to make any actual use of social media in a crisis situation of any size there needs to be some level of automation to assist in the process. In this section, we have looked at the characteristics of social media messages with a focus on tweets and discussed where a filtering mechanism would be useful to support more effective information gathering for crisis management by differentiating between information types and sources of messages generated during a crisis.

While focusing on which types of social media communications to process will help to reduce the problem of processing a huge volume of information, there are other problems. In the next section, our discussion focuses on handling multiple languages in social media communications during a crisis in which more than one language is used by eyewitnesses and authorities.

8.4   Multiple Natural (Human) Languages in Crises

The attacks in Brussels, particularly the attack on the airport, highlight one very important issue in such cases: the world of social media is not monolingual.

Some crises are of such a nature and take place in a geographical area in which one language is sufficiently dominant to be the only language needed for gathering social media information to support situational awareness. An example of such a localized crisis would be the severe flooding which took place in northern England at the end of December in 2015. Even without examining the collected tweets or Facebook postings from that event, it would not be unreasonable to assume that the vast majority of social media communications from the populace on the development of events were written in English.

However, the March 2016 attacks in Brussels are at the opposite end of the spectrum. As mentioned previously, Belgium has three official languages: French, Dutch (Flemish), and German. Even just limiting communications to those of Belgian nationals whose first language is one of these three, the ability to retrieve actionable information from social media generated by them would require the ability to process and synthesize information presented in all three languages.

Additionally, due to its position as the seat of the European Union, and its large expatriate communities composed of immigrants, EU workers, lobbyists from a variety of organizations and countries, as well as personnel of international corporations with their European headquarters in Belgium, there are a very large number of other languages which are routinely used in the capital. Complicating the issue even further in the Brussels attack was that one of the targets was the international airport on the northern outskirts of the city, where victims and witnesses included arriving, departing, and transferring passengers from a multitude of countries. Tweets and other communications from the injured and traumatized observers and victims would mostly have appeared in their mother tongues. Thus, if a social media scanning system for crisis management is only able to deal with a single language, then many important puzzle pieces could be well inaccessible, and the situational picture is the poorer for it.

Furthermore, in situations such as in Haiti in 2010, extracting actionable information during a large-scale natural disaster such as a major earthquake or tsunami may depend on the ability to deal with natural language issues due to the presence of external players such as international aid organizations whose working languages are not only not always that of the local population, but often not the same as the other agencies which they need to be able to communicate and coordinate with during the emergency. This means that each actor within the crisis area may well have some important information that is not accessible to other.

However, multi-linguality comes at a cost. Processing power is needed to translate from language to language, which may be easily stretched very thin during a time at which a huge volume of time sensitive information is being generated (Fig. 8.1).

Addition of a new language (n + 1) results in the need for n new translation connections, a hefty undertaking.

Therefore, it obliges us to look at shallow, less processing intensive methodologies which may allow us to quickly access information in various languages; our solution is described below.

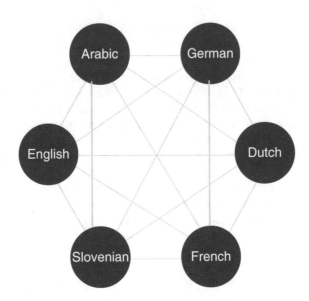

Fig. 8.1 Translation from one language to another requires (n − 1)! translation engines

8.5 Text Analytics

There are many different approaches to text analytics, which perform various functions such as sentiment analysis based upon the presence of keywords showing
emotions such as 'afraid,' 'angry,' 'worried,' or 'shocked.' Others simply count the
occurrences of words or phrases within the corpus to identify trending topics. Often
the results of such scans are presented as word clouds, which provide crisis managers with an overview of topics being discussed. Although such analyses provide no
concrete actionable information, they may be useful in supporting crisis managers
with the development of public information announcements to calm fears and provide useful information during chaotic times.

Other text analytic tools can be useful to extract certain types of information
from natural language documents, such as identifying named entities (i.e., persons,
places, organizations, etc.) or patterns such as hashtags, telephone numbers,
addresses, and so on. More complex processing may allow for the extraction relationships between entities that can be useful for decision-makers.

In general, more complex text analytics is needed to do translation from one
language to another. Words in one language seldom map one-to-one into another
language, meaning the clues picked up from context determine the meaning of the
word. For example, the English word 'train' may be a noun which refers to a mode
of transportation on rails, a verb which has to do with teaching someone to perform
a skilled action, or a long dragging hemline on a woman's wedding gown, to name
just a few. In order to decide which role 'train' plays in the sentence, an analysis
must be performed which looks at other structures in the sentence and derive from
them the most likely corresponding word in the target language.

However, there are some characteristics of social media to keep in mind. One of the important ones, from a text analytic point of view, is that message style tends to be informal from the grammatical point of view, with fragments rather than complete sentences. This means that markers such as capitalization or punctuation upon which text analytic programs rely may be missing. For example, in English, one indicator of a named entity (proper noun), such as a person, location, etc., is that the first letters are capitalized, whereas common nouns are uncapitalized. When capitalization is not used, it is more difficult to identify the named entity. (It may still be possible from analysis of surrounding text which may provide disambiguating context, but the process requires more intensive processing.) Additionally, in many languages, the normal rules of grammar fall by the wayside in texting, particular when the texter is stressed, injured, or under shock. For example, in German, nouns are generally capitalized, whether proper nouns or not; this rule is often relaxed in social media. Likewise, influences such as the 140-character limit in Twitter have led to truncated or fragmented sentences, as well as abbreviations such as lol, cu, and imho, as well as nonstandard spellings (e.g., gr8 for "great"), which the underlying lexica need to accommodate. Under stress, even those minimalized remaining grammatical rules may no longer hold as niceties are ignored for expediency. Likewise, under stress misspellings are much more common.

Therefore, many of the rules upon which some of analysis for translation is based may simply not function in a crisis situation.

8.6 The Problem of Synonymy

As mentioned in the previous subsection, mapping of words from one language to another is seldom one-to-one. Additionally, there is the issue of synonymy within a single language, that is, the ability to describe the same event or state using different formulations. For example, English-speaking humans recognize quite easily that 'exploded,' 'blew up,' 'detonated,' and 'fireball' are various ways of describing the same thing (an explosion), and thus would be able to cluster messages containing these expressions into the same category.

Synonymy plays a very important role if we are filtering the deluge of streaming information looking for a specific type of event or object. Humans generally can quickly decide if a particular word or phrase may be descriptive of that which is being searched for, as shown above. However, a computer program can do this only when it has the information that these are all synonyms.

Therefore, identifying which communications relate to the incident we are interested in, as well as determining which belong together (relate to the same incident, as opposed to separate incidents), and culling out those which are unrelated generally requires more than simple word matching.

In the following section, we look at a solution which was developed for use within the ATHENA project for multilingual crisis communications which reduces translation overhead by exploiting a standardized representational language to reduce the overhead of translation.

Normally, in translating from one language to another, the translator attempts to adopt the style in which the original text was written for the target language, that is, if the original formulation was written casually, one tries to avoid a more formal style in the target language. However, within a crisis situation, we are far more interested in the communication of pertinent information than we are in the stylistic aspects of the formulation of that information.

Within the ATHENA project, we examined a shallow approach to support the use of multiple languages. This approach is based upon the use of crisis management language, which we will describe in the following section.

8.7   Crisis Management Language

Crisis Management Language (CML) is a subset of a NATO R&D initiative which originated out of the area of modelling and simulation. In the military domain, Battle Management Language (BML) has been developed to formalize command and control communications including orders, requests, and reports [5, 6] between command and control systems of various nations in various languages. The use of such a controlled, formalized language enhances interoperability and prevents mis-interpretations. In addition, all expressions are processable by systems as well as human-readable. Obviously, the BML technique also can be applied for C2 communication in disaster relief operations in order to upgrade the coordination and the collaboration of the many participating organizations and their relief units. Since military units are very often involved in peace-keeping, disaster relief, and other nonwar operations, the basic structures for the development of CML already existed within BML and only needed to be expanded for C2 communication in disaster relief operations in order to upgrade the coordination and the collaboration of the many participating organizations and their relief units.

C2LG is the formal grammar that is used to generate expressions of military communication (orders, requests, and reports). The resulting language, that is the set of all expressions that can be generated out of the C2LG, is called Battle Management Language (BML). Since C2LG is a formal grammar, the expressions generated can be analyzed (parsed and interpreted) by automated systems. It can be used to exchange expressions between the C2 systems of allied forces or to exchange messages between a C2 system and a simulation system in order to use the simulation system for decision support or training. C2LG uses a lexicon of terms well defined in the field of military operations to generate standardized, unambiguous, human- and machine-readable expressions. C2LG exploits the idea of the 5 Ws, meaning that an expression consists of Who, What, Where, When, and Why [5, 6]. As such it can be used to support multi-agency operations in the domain of crisis management [7, 8].

CML provides a means to express directives (orders and requests in the case of military operations, and tasking and requests in the case of civil operations) and reports in a formalized and unambiguous way. Incoming natural language (text) information can be processed and reduced into a controlled CML statement to enable sense-making algorithms to identify hotspots, eliminate ambiguity, and assess the situation. Furthermore, communications between responders from different agencies may be formulated using CML to eliminate ambiguity and reduce errors. Additionally, C2 systems (police, fire, etc.) may have a CML-GUI integrated into their operations which can be used for the formulation of communications that conform to CML. The GUI allows the formulation of the directives and reports to be communicated. Operating the GUI is very intuitive because it is similar to natural language. This reduces the learning curve for users and ensures communications are correct even in stressful situations. The GUI enforces the use of specific terms and also enforces a restricted grammar. In addition, it provides a window that shows on a map the positions of units (e.g., ambulances), facilities (e.g., hospitals), the places where disaster has struck and where patients wait to be picked up. This window provides situational awareness as it is updated according to incoming reports. In summary, the GUI ensures the compliance of the formulated expressions with the CML standard and sends them to the selected addressee(s) among the participating units and organizations. When a CML message is delivered to its addressee, it automatically appears in the addressee's GUI.

Other features have already been added to CML such as the ability to express a first diagnosis. This first diagnosis normally is provided by the emergency physician who treats the patient on-site. The availability of the first diagnosis via CML helps the control center to offer the best hospital for the patient in question even faster. In addition, all information can be forwarded to the chosen hospital to prepare emergency treatment in time. The hospital also profits from the fact that the ambulances report their current position in regular time intervals whose length can be set as necessary. Thus, a hospital has the necessary information about patients coming in, information about the injuries, and information about the point in time at which the patient will arrive. In the case of a disaster relief operation, when there might be many more patients to be treated in hospitals than on any normal day, the information advantage provided to hospitals by the use of CML can play a critical role in saving lives. This is even enhanced by the fact that the control center is supported in its coordinating function.

The grammar of CML uses formal context-free rules for building messages. The order of the elements as well as keywords is determined by the production rules, producing statements which are both machine-processable and human-readable, as shown in the example below:

- **report event** PoliceUnit13 gas-explosion at building-1312 ongoing at 201111051413 explosion_report_01;
- **request rescue** PoliceUnit13 C2RFS one injured civilian at building-1312 start asap rescue_request_02;
- **request support** Ambulance2 C2RFS rescue_task04 by crane request_support_06;

- **request support** C2RFS C2_MIL rescue_task04 by crane request_support_07;

Each CML statement begins with a keyword indicating the time of communication ('request,' 'report,' etc.) followed by a standardized, unambiguous 'verb' which identifies the substance of the communication ('event,' 'rescue,' etc.). Each following constituent provides more detail in a standardized way. If desired, the receiving C2 system converts the English-like CML statement into the natural language of that C2 system, i.e., a French C2 system would do a constituent by constituent conversion into French, so that the operator at the C2 system would not be required to understand English.

8.8 'CML Light' for Social Media in Crises

As the reader may have noted, the 'full version' of CML is designed to provide complete information in each statement. But as we have seen earlier in this chapter, social media presents diverse challenges, among which are fragmented and truncated communications. However, some of the principles underlying CML are still useful for our purposes in a 'light' version.

CML contains standardized expressions for specific events, objects, and activities. The original core of CML was derived from the Joint Command and Control Computer Information Exchange Data Model (JC3IEDM) [9], a data model used for interoperability in NATO. The terms and their meanings within the data model were selected and agreed to by 28 nations, reflecting a common understanding of each term.

For CML, we have exploited this common terminology and have used the terms originally contained in BML where they exist. Where there were none already available, we have expanded CML's lexicon with standardized terminology for specific events, states, or actions within the domain of crisis management.

Through this, the issue of synonymy as discussed above is resolved. CML contains an event called 'explosion.' There are many words in English to describe an 'explosion' such as 'exploded,' 'blew up,' 'sounded like a bomb,' 'kaboom,' etc., all of which, when found, map to the single JC3IEDM 'explosion' (Fig. 8.2).

Similarly, other languages have their own set of synonymic expressions—some languages more variations than English, some less. But we can still map those variants to the same expression. For example, if we add German to the mix, we can similarly map to the standardized CML 'explosion' (Fig. 8.3).

Because CML recognizes the synonyms as indicators of the same event, we can, for example, scan German social media messages, and recognize these words and phrases as indicative of an explosion, thus filtering out the interesting messages based upon a fairly simple process.

Furthermore, by mapping onto a single, central term, we reduce the complexity of the translation process dramatically, as we do not try to find stylistically similar forms in the other language.

Another advantage is demonstrated in Fig. 8.4. Adding capabilities for processing messages using a new language, we need only identify synonymic words and phrases and map them onto a single term.

As a result, we can reduce the complexity of the translation process dramatically. Whereas in Fig. 8.1 earlier in the chapter we would have needed $(n - 1)!$ translation engines for n languages, we now need only n engines, as demonstrated in Fig. 8.5.

Furthermore, the addition of a new language to the mix simply requires the mapping of terms and synonymic expressions of that language into the CML vocabulary, a dramatic reduction in complexity and effort.

By using this modified 'CML light,' in which we are focusing on vocabulary rather than the full set of production rules, we can quickly identify which social media messages are of immediate interest and which can be ignored or wait for delayed processing by simply scanning streaming information looking for patterns which map to events, states, or actions that we are interested in.

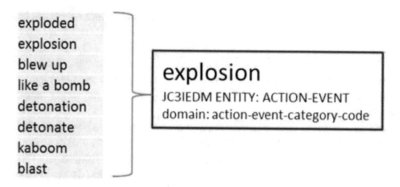

Fig. 8.2 Mapping of English synonymic expressions indicating an explosion to a standardized expression

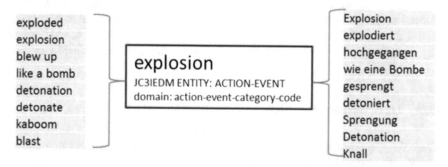

Fig. 8.3 Mapping of English and German synonymic expressions indicating an explosion to a standardized expression

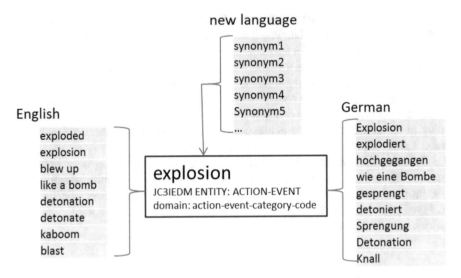

Fig. 8.4 Adding a new language simply requires mapping of its synonymic expressions indicating an explosion to a standardized expression

We can further exploit this for multilingual audiences in that, for each list of synonymic expressions, one of these will be designated as the unambiguous expression for translation out of the CML vocabulary into the target language as shown in Fig. 8.6.

So, if we get a message in Slovenian that three people have been hurt, CML will produce a message *wounded 3 civilians*, and the German users will get the message '3 Personen verletzt' and the English users will get '3 people injured'... (The 'word order' of CML is designed for faster computer processing but is not 'natural' for humans, so we do some minor conversions.)

Thus, translation from one language to another can be accomplished quickly, even if style and grammatical correctness suffer. However, humans are relatively flexible and can fairly easily fill in the gaps, even with a shorthand version of information. Having information available to all relevant players, regardless of which natural language they use, solves many problems. However, the greatest advantage is that this further reduces the overhead for running automatic filtering and clustering algorithms, allowing for more timely updates to the shared situational picture.

8.9 Information Quality

Once the above issues have been dealt with, there are still the problems of unintentional misinformation and intentional disinformation. Unintentional misinformation is when incorrect information is disseminated, but without malicious intent. This may cause many problems for responders by tying up unnecessary resources when incorrect information comes in. This can only be countered by comparing

Fig. 8.5 CML in the middle reduces the complexity of translation

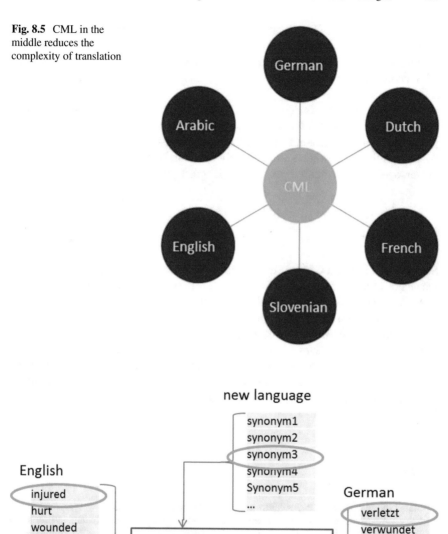

Fig. 8.6 For each language, an unambiguous term will represent its synonymic cluster

information from known, reliable sources to that from unknown sources. This may, however, in international situation be alleviated because the reporting from all known reliable players such as emergency or government sources will be available to all regardless of the language that information was generated in.

Disinformation, in which intentionally misleading or incorrect information is disseminated through social media by individuals with the intention to disrupt ongoing operations, may be much more difficult to detect. Again, however, cross-checking with confirmed information from known sources may identify disinformation sources more quickly.

The issues of both misinformation and disinformation in crises may be ameliorated by algorithms which can operate upon information standardized through a process such as conversion to CML. Such algorithms are able to sift through and compare information much more quickly than humans, in particular when those human operators and analysts are also under great pressure.

8.10 Summary

Social media can be a very powerful support mechanism for emergency responders, managers, and the public in crisis situations. However, the use of it in such situations depends on the ability to deal with various issues such things as volume, pertinence, and natural language. In this chapter, we have discussed a number of issues surrounding the use of social media with a focus on large international crisis situations. It is thus very clear that in order to make any actual use of social media in a crisis situation of any size there needs to be some level of automation to assist in the process. We have shown a solution which has been developed within the Athena project, which we believe solves many of the communications problems found in large-scale, international crises.

References

1. http://edition.cnn.com/2013/04/23/tech/social-media/tweet-ripple-effect/.
2. Siegel, A. (November 16, 2015). Here's what we can learn from how Twitter responded to Paris. *Washington Post.* Monkey Cage blog. Retrieved from www.washingtonpost.com/news/monkey-cage/wp/201/11/16/heres-what-we-can-learn-from-how-twitter-responded-to-paris.
3. Terpstra, T., DeVries, A., Stornkman, R., & Paradies, G. L. (2012). Towards a realtime Twitter analysis during crises for operational crisis management. In O. Rothkrantz, J. Ristvej, & Z. Franco (Eds.), *Proceedings of the 9th International ISCRAM Conference*, Vancouver, Canada.
4. Olteanu, A., Vieweg, S., & Castillo, C. (2015). What to expect when the unexpected happens: Social media communications across crises. In *Proceeding from the 18th ACM Computer Supported Cooperative Work and Social Computing, CSCW'15*, Vancouver, Canada. Retrieved from http://crisislex.org/papers/cscw2015_transversal_study.pdf.
5. Schade, U., & Hieb, M. R. (2006). Development of formal grammars to support coalition command and control: A battle management language for orders, requests, and reports. In *11th ICCRTS*, Cambridge, UK.
6. Schade, U., Hieb, M., Frey, M., & Rein, K. (2010). *Command and Control Lexical Grammar (C2LG) specification*. Version 1.3. FKIE Technical Report.

7. Remmersmann, T., Rein, K., & Schade, U. (2011) Coordinating ambulance operations, Future security 2011. In J. Ender (Ed.), *Proceedings from the 6th Security Research Conference*, Berlin, Germany (pp. 47–50). Stuttgart: Fraunhofer.
8. Coote, R., Rein, K., Esch, M., & Schade, U. (2015). Towards a mobile context-sensitive framework for interoperability and improved situational awareness in crisis and emergency management. In J. Beyerer, A. Meissner, & J. Geisler (Eds.), *Proceedings from the 10th Future Security 2015, Security Research Conference*, Berlin, Germany. Stuttgart: Fraunhofer.
9. JC3IEDM. https://mipsite.lsec.dnd.ca/Public%20Document%20Library/Forms/AllItems. aspx?RootFolder=%2FPublic%20Document%20Library%2F03-Baseline_ 3%2E0%2FJC3IEDM-Joint_C3_Information_Exchange_Data_Model.

Chapter 9
Cloud-Based Intelligence Aquisition and Processing for Crisis Management

Patrick de Oude, Gregor Pavlin, Thomas Quillinan, Julij Jeraj, and Abdelhaq Abouhafc

9.1 Introduction

Contemporary crisis management has to deal with complex socio-technical environments involving many interdependent elements. Many dependencies in such settings result in complex cascading processes that can have adverse effects on the population, environment, and economy. Adequate situation awareness and the capability to predict the development of a crisis situation under different circumstances is a critical element of effective, and timely, crisis management and response. However, this requires rich domain knowledge, as complex interdependencies between different socio-technical elements must be understood. Furthermore, substantial domain knowledge is required for (1) determination of what data is relevant in a given situation, (2) the collection of that data, and (3) its analysis; i.e. all the information should be delivered to an expert that can understand that information. Note that domain knowledge is required to 'drive' the information requests. Moreover, often multiple experts from different organizations need to be involved, as a single person cannot understand all aspects of the crisis, nor can he/she process all the relevant data. This implies delegation of work in cascaded collaborative systems (distributed awareness), where each expert works on a subset of the problem. The experts must collaborate to supply the decision makers the right information such that they can manage the crisis effectively. To do this effectively all the relevant data must be supplied to the appropriate experts on time and, at the same time,

P. de Oude (✉) • G. Pavlin • T. Quillinan • A. Abouhafc
Thales Research and Technology Netherlands, Delft, The Netherlands
e-mail: Patrick.deoude@d-cis.nl; Gregor.pavlin@d-cis.nl; Thomas.quillinan@d-cis.nl; Abdelhaq.abouhafc@d-cis.nl

J. Jeraj
City of Ljubljana, Emergency Management Department, Ljubljana, Slovenia
e-mail: Julij.jeraj@ljubljana.si

© Springer International Publishing AG 2017
B. Akhgar et al. (eds.), *Application of Social Media in Crisis Management*,
Transactions on Computational Science and Computational Intelligence,
DOI 10.1007/978-3-319-52419-1_9

information overload must be avoided. The data is relevant with respect to the task an expert is performing at a given time.

In this chapter, we present the A-Cloud and the ATHENA Logic Cloud (ALC) approach that addresses the above-mentioned challenges. The ALC facilitates efficient creation of collaborative decision support solutions for complex crisis management processes. The ALC solution supports a cost-efficient, seamless linking of 'deep expertise' provided by humans or automated solutions, and social media. Such seamless combination of data sources and analysis capabilities is a critical capability required for informed and timely crisis management. This chapter focuses on two types of technologies enabling advanced situation awareness: (1) automated information management based on inter-operable services, and (2) data-bound security. Overall, the ALC solution allows cost-efficient implementation of technical platforms enabling rapid exploitation of rich domain expertise, thus improving the quality, coverage, and speed of sense-making processes (see Chap. 3). The challenges and solutions are illustrated with the help of a running example involving crisis management in a case of large-scale water pollution.

The ALC was demonstrated in a live testing exercise in Ljubljana, Slovenia, where different stakeholder organizations involved in crisis management response participated, such as, Municipal Chemical Crisis Manager, Municipal Water Company, National Environmental Agency, Municipal Fire Department (fire fighters), and citizens (see Chap. 12). These different stakeholder organizations need to do collaborative sense making in case of the large-scale water pollution incident that could potentially have a devastating impact on the aquifer in the surroundings of the city of Ljubljana, as the ground water represents the main source of fresh water.

This document is organized as follows: the first section provides a problem definition illustrated with the help of a real world scenario. The second section introduces the A-Cloud and the main principles of the ALC. In the following section, the security issues of the A-Cloud are discussed. In Sect. 9.6, we describe how the ALC was applied in the ATHENA Ljubljana live test exercise (see Chap. 12). In the last section, the chapter is summarized and some conclusions are drawn.

9.2 Problem Definition

This chapter is addressing contemporary crisis management problems in complex socio-technical environments characterized by many interdependent natural and man-made processes, resulting in chains of cascading events [1] that can have adverse effects on the population, environment, and the economy. Achieving and maintaining adequate situational awareness in such environments is a non-trivial challenge that can be characterized as follows:

- Rich expertise/domain knowledge is required to (1) understand the collected data and (2) be able to find and access the right data (types and context); an expert should be provided only the information he/she can understand. Such domain knowledge 'drives' information requests,

- Multiple experts from different organizations have to be involved → delegation of work in cascaded collaborative systems (distributed awareness, where each expert works on a subset of the problem),
- All experts cannot meet face to face → work remotely,
- Supply all the relevant data to the right experts on time and, at the same time, avoid information overload. The data is relevant with respect to the task an expert is performing at a given time,
- Supply the experts at different levels of the collaborative hierarchical system with the right types of data from social/new media,
- Supply command and control systems with the outputs from the right experts from the different levels of the collaborative hierarchical system, and
- Experts must collaborate to provide decision makers with the right information at the right moment in time such that the crisis can be managed effectively.

In other words, the targeted crisis management applications involve Communities of Practice sharing various types of knowledge [2]. Particularly relevant challenge is Knowledge Brokerage.

In this chapter the challenges and solutions are described with the help of a realistic scenario that was used at the ATHENA live testing exercise in Ljubljana, Slovenia (see Chaps. 12 and 13). One part of the exercise focused on assessing the situation and the impact of potential actions in case of a chemical incident. An incident at a chemical plant close to Ljubljana resulted in a substantial release of a mixture of styrene and gasoline into the Sava River (see Fig. 9.1). Up to 5 tons of these pollutants might have escaped, which could have had a devastating impact on the aquifer in the surroundings of the city of Ljubljana, as the ground water represents the main source of fresh water.

In the presented scenario, the risk of severe water well pollution is high, as the river flows through an area with porous fluvial sediments that is likely to facilitate the penetration of the pollutant. The area is the main source of the fresh water provided by the water company. If the water discharge in the river were increased, the pollutant would quickly be transported past the porous river banks making the impact negligible. However, while simple and locally effective, such a measure might have a negative impact on a larger area downstream. Obviously, the simple decision about the action requires the knowledge about the current situation and a lot of domain knowledge and analysis capabilities combined with political negotiations, such as:

- the type and the quantity of escaped chemicals;
- the extent of the river pollution;
- the impact of the chemical on the health, in case the ground water is polluted.
- the impact of the chemical on the river itself (animals, plants);
- the propagation of pollutants through the geological layers (there exist maps capturing the types of rock formations and the speed of the ground water); and
- the possible measures (controlling the water flow in the river, barriers, etc.).

Fig. 9.1 Pollution downstream Medvode (this image is courtesy of http://akvamarin.geo-zs.si/
incomepregledovalnik/)

Such knowledge and capabilities are not supported by a single stakeholder. Instead, different professionals, each with a specific expertise and analysis capabilities must collaborate to solve the problem; i.e. the professionals establish distributed awareness systems where each expert provides specific services and communicates with other stakeholders only about the information relevant for a specific task. The resulting systems can be viewed as a composition of heterogeneous services connected via dedicated information flows. The information flows must be created dynamically, as the demand for specific services arises. Figure 9.2 shows an example of such a collaborative system. In the presented example the following services are used.

Municipal Chemical Crisis Manager (CCM). (Member of the city's Civil Protection HQ staff, and an employee of Municipal Emergency management department): collects all the relevant information regarding the environmental aspects in the crisis, such as analysis reports from Water Companies, analysis reports provided by the Environmental Agency as well as selected reports from citizens. The CCM prepares reports (1) on the areas in which the water is not potable or (2) how the water pollution will evolve in the near future. The recipients of the CCM's reports are coordinators in the municipal government and Command & Control (C2) operators.

Municipal Water Company (MWC). Operates the internal company's ground water monitoring system and establishes emergency ground water quality control within the company (tap water, water wells, piezometers). The MWC periodically collects all the relevant information on the wells and piezometers from within the

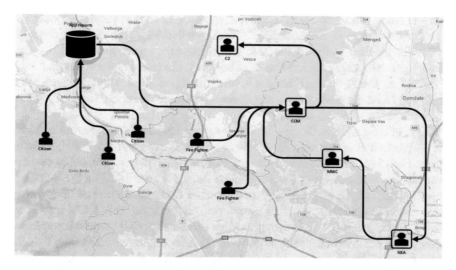

Fig. 9.2 Collaborative situation assessment involving different stakeholder organizations in a large-scale water pollution incident

organization and prepares a sequence of reports and prognosis of the places where the water is not potable and/or might not be potable within certain time periods. The MWC also declares water potable/non-potable and introduces rules on water consumption. The MWC also requests and analyses data from other ground water monitoring providing entities, such as the National Environment Agency, the city's environment department, and contracted companies. The CCM is recipient of the MWC reports.

National Environmental Agency (NEA). Operates national ground water and surface water monitoring system, provides data on river water levels, discharge, velocity, maintains rivers, cleans rivers after pollution, estimates the current extent of the pollution, and predicts the evolution collects all the relevant information on (1) the wells from the water companies, (2) surveillance of the surface water provided by trained fire fighters, and (3) measurements by specialized environmental agency experts/mobile labs. The NEA has deep understanding of the physical aspects of the aquifer. Recipients of the NEA's reports are national environment authorities, national emergency management authorities, and the CCM. The NEA uses measurements from the Water Company, specialized agency experts/mobile labs, and reports from fire fighters.

Municipal Fire Department (Fire Fighters). Responds to chemical release in the chemical plant by sealing off the rupture and preventing further release to the river. The MFD also provides professional monitoring service for the area, such as (1) obtaining data on released chemicals from the company and/or demanding that service from national emergency management authorities, (2) identifying chemicals, (3) obtaining data on river water levels, discharge, velocity from environmental agency, (4) providing professional monitoring of the surface water; can recognize certain types of pollution (e.g. chemicals, oil, sludge), and (5) decides on where to put booms, skimmers, sorbents, and removes chemicals from the river. The observations are reported to the CCM. This service requires substantial knowledge and training to recognize pollutants, and estimation of the extent of the pollution.

Citizens. Use Apps and other social media to report about pollution and receive info and recommendations/instructions from authorities via social media and other type of media.

One of the major challenges in domains involving such heterogeneous services is the creation of information flows between the right experts at the right moment in time. Each expert requires timely delivery of specific types of information that he/she can process. Moreover, the information flows are established on the fly as the needs arise. In such a system an expert searches for specific types of information from a certain context (e.g. area, time interval, etc.). For example, CCM is looking for observations of pollution symptoms in a certain area. CCM would formulate a query that should automatically be delivered to the specialized teams of fire fighters who can provide the service. After teams of fire fighters reaches the specified area and obtains the requested data, the messages should be automatically returned to the CCM. In other words, the collaboration depends on dynamic creation of information flows between the services consuming and producing the data, respectively. Note that this can be asynchronous querying of services that can produce the data of interest. Such a collaborative approach to data acquisition and processing, however, is difficult to achieve with the standard communication and collaboration approaches (e.g. phones, emails, web portals). The following section describes the ATHENA Logic Cloud, a technical solution that facilitates implementation of such collaborative crisis management solutions.

9.3   ATHENA Logic Cloud

To facilitate automated information management between different stakeholder organizations during large-scale crisis event, we introduce the ATHENA cloud (A-cloud) solution. The A-Cloud consists of two main components (see Fig. 9.3), namely: (1) the ATHENA Persistence Cloud (APC) that provides storage capacity that can be based on an arbitrary mixture of technologies, such as heterogeneous databases, shared distributed data-spaces (DDS), etc. and (2) the ATHENA Logic Cloud (ALC), based on the Dynamic Process Integration Framework (DPIF) [3, 4], that provides an abstraction of the accessing and manipulation processes for the data contained in the APC or other data sources and the Martello security layer, that provides an encryption service, allowing secure operation in open networks by providing data-bound security solutions. (More details regarding Martello are discussed in Sect. 9.5.) Moreover, the ALC allows targeted querying of sensors and observers at runtime (e.g. request pictures from a certain camera; ask LEA agents for a report from a specific location, etc.). In other words, ALC introduces automated Knowledge Brokerage between the different types of services. Note that, the

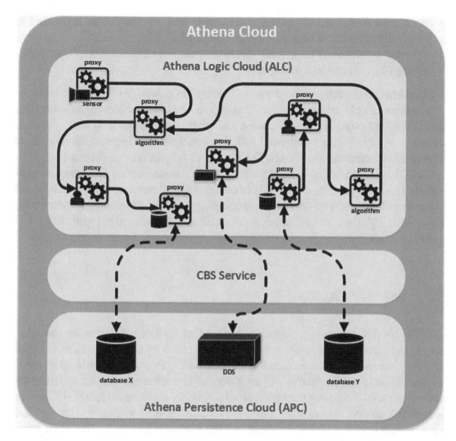

Fig. 9.3 The A-Cloud with the ATHENA Logic Cloud (ALC) and the ATHENA Persistence Cloud (APC) components. The *black unidirectional arrows* represent information flows between different service proxies in the ALC, while the bidirectional, *dashed arrows* represent interfaces to different data sources, such as databases and distributed data spaces in the APC. The CBS service will be addressed in Sect. 9.4

ALC will not store any data: it is a 'process' cloud, introducing the interoperability and the logic for the creation of information flows between data consumers and providers, on top of the persistence cloud.

The ALC is a *community cloud* that supports access to different systems and services provided by a group of organizations. Providing systems and services through a cloud environment provides numerous advantages over other types of solutions, namely:

- cloud computing provides centralized storage and online access to computer resources and services → resource sharing;
- resources are provided over the network in a manner that provides platform independent access;
- reliability is increased due to use of redundancy'

- maintenance of cloud computing applications is easier, since there is no need to install the application on a user's computer; and
- costs can significantly be reduced due to the flexibility in use, maintenance, monitoring of cloud services and systems.

The ALC community cloud supports Platform as a Service (PaaS) and Software as a Service (SaaS) service models. By using the ALC, stakeholder organization can provide their system services in the cloud platform environment, while domain experts can directly use a dedicated software graphical user interface to share their expertise with other organizations. The ALC is based on a service-oriented architecture (SOA) platform supporting loose coupling of resources from different data providers. This is an important capability in domains where creation of a single database space (e.g. integration of data in a centralized system) is technically, economically, and politically not a viable solution. We will start the discussion with the DPIF proxy concept.

9.3.1 Service Proxies

A DPIF service proxy is a software process that continuously runs on the ALC server and represents a specific service provided either by a domain expert or by an arbitrary automated solution (see Fig. 9.3). A DPIF proxy can be seen as an automated 'Knowledge Worker' enabling Knowledge Brokerage [2]. An expert communicates with the proxy through a suitable human-machine interface (HMI). An automated service, on the other hand, can be implemented within the proxy or run on a remote device accessed by the proxy via a secure connection. The most common types of automated services are:

- Sensor drivers and web service clients that provide access to the sensor data and control parameters. Examples of sensors that can be integrated are cameras, thermometers, motion detectors, etc.
- Data store access mechanisms that allow manipulation of different types of databases, distributed data stores, etc. The code implementing web service calls, database connectors, and other access methods is typically contained within a proxy.
- Arbitrarily complex analysis processes (algorithms) supporting inference about correlated data (e.g. classification, prediction), data mining, machine learning, etc. A proxy can either encapsulate an algorithm or interface the algorithm via a suitable API, often implemented through web services. The latter is the case if automated processes require substantial computing resources and/or belong to legacy systems (e.g. stove-pipes).

Overall, a service proxy makes a specific service interoperable and provides the logic enabling service discovery and dynamic creation of information flows between complementary services. In other words: interfacing between services is 'outsourced' to the proxies in the ALC, each proxy provides an interface between a very specific process and a standardized platform implemented by the ALC. Each proxy publishes the service it represents, and subscribes to complementary services in case additional inputs are needed. The discovery of the right services is based on a combination of service directories (yellow pages) and negotiation. In this manner, a consumer is connected with the right services providing the right type of information from the right context.[1]

9.3.2 Interoperability: Principles and Tools

Collaborative analysis processes critically depend on interoperability between complementary services sharing different types of information. Each proxy representing a service adopts suitable standards and translates specific service outputs to standardized data objects that can be shared with proxies representing consumer services. The standards define the semantics and the formats, in which the data is exchanged. In ALC a service standard specifies multiple elements:

- The semantics and formats of the service type including the context,
- The semantics and formats of the data produced by certain type of complementary services (inputs), and
- The semantics and formats of the service invocation inputs, required to activate the service.

In ALC, a service standard consists of: (1) free text describing the service, (2) the service invocation data structure, and (3) the service response data structure [4]. Note that the service invocation and response data structures can be arbitrarily complex data objects defined through multiple attributes of different data types. For example, the service response of a monitoring service provided by a fire fighter consists of attributes, such as a text field, an image, a map object (for the specification of the location at which the report was created) and the response time.

The targeted domains, however, are characterized through a great variety of service types. Moreover, new services are added at a fast pace. For example, a new expertise on handling a certain type of chemicals might be acquired by a participating organization. The new service type definition has to be introduced and advertised in such a way that potential consumers can easily understand the service, activate it and use its outputs.

An important feature of DPIF is the ease of adopting the existing and creating new service standards. The key enablers of this are the design of the service proxies in combination with OntoWizard, a special configuration tool [4]. By using the OntoWizard, a service provider can describe the provided service and the required types of inputs in a standardized way in a few simple steps, without requiring any technical knowledge of the underlying DPIF technology and service representations.

---

[1] In DPIF context can be a simple location, or an arbitrarily complex combination of attributes, such as location, time interval, service quality, etc.

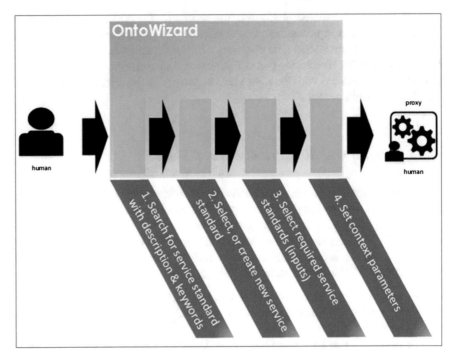

Fig. 9.4 Steps of the OntoWizard configuration tool to add a human-based service (e.g. human observer, analyst, data consumer, etc.) to the ALC

The configuration process produces rigorous metadata used by the proxy representing that service. The metadata is used for: (1) the publication of the service, (2) the subscription to outputs of other types of services that the service requires, and (3) negotiation that filters adequate services based on the captured context.

The common type of services that were discussed in the previous section, 'Service Proxies', can be categorized as human- and machine-based services. The process to incorporate human-based services into the ALC environment is depicted in Fig. 9.4 describing the following steps:

1. Finding the right service standard: the user of the proxy (i.e. the service provider, such as for example an aquifer expert) uses a special service browser tool. The expert first describes the service through free text that is used by the tool to search in the catalog containing the standards for services that have already been implemented in ALC. The tool returns a ranked list of service standard descriptions that facilitates inspection by the expert,
2. Service type selection: if expert has found a suitable service standard describing his/her capability in the ranked list of service descriptions, the expert simply selects that service resulting in automated creation of rigorous metadata that is used by the proxies in ALC for service discovery, filtering and creation of information flows. However, in case no match is found, a new service can be created and added to the catalog. The expert can use a special editor to formulate

a new service standard that describes the new capability. The new service description is added to the service standards catalog and can easily be adopted by other experts providing the same type of a service or consumers,[2]

3. Subscribing to complementary services: to be able to provide a specific service, an expert might require output of other types of services provided by other proxies. The OntoWizard tool allows the specification of such complementary service types. By using free text description of the required service types, the expert can query the catalog. The system returns a ranked list facilitating the inspection of potentially suitable service types in ALC. Upon selecting suitable service types, the system automatically creates metadata used by the proxy to subscribe to the right types of complementary services.

4. Setting the context: the service provider uses OntoWizard to specify the context of the provided service. The context parameters specify the conditions under which the service can be provided. For example, geographical location, time intervals in which the service is available, quality of service, clearance, etc.

In summary, by guiding the service providers through a sequence of simple configuration steps, the OntoWizard tool creates rigorous metadata that the proxies in ALC use for the automated creation of information flows supporting the delivery of the right information to the right stakeholders at the right moment in time.

By using the same tool, machine-based services can be introduced into the ALC. The process of creating a machine-based service proxy is illustrated in Fig. 9.5. The process is almost identical to the human-based service creation of a proxy except two additional steps need to be performed prior to using the OntoWizard tool as shown in Fig. 9.5. These steps involve programming the code of the proxy that interfaces with the database (using an Integrated Development Environment (IDE), for example). Such code typically incorporates the logic that supports interaction with the database and extraction of information. For more details about the machine-based creation of proxies consult [5].

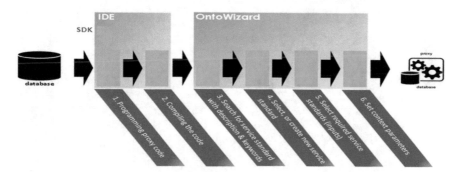

Fig. 9.5 The process of creating a proxy for a database

---

[2]The creation of new services will require some form of governance to ensure effective use of the defined service standards.

The OntoWizard tool and SDK are used by all users/developers contributing various services and provide a standardized way of creating service standards. In other words, DPIF powered platforms, such as ALC, are open ended regarding the standards; as new services are introduced, the corresponding service standards can be introduced on the fly. Note, the DPIF SDK and OntoWizard tool standardize the creation of service types and service proxies.

9.3.3 Contextualized Request and Delivery of Information

Data that is requested or delivered is usually created in a specific context (e.g. area, time interval, clearance). Consequently, the information flows between proxies are always associated with a certain context, i.e. conditions. To illustrate this, consider the example in Fig. 9.6 that focuses on the information flow between the fire fighters and the CCM discussed in Sect. 9.2. In this particular situation, the CCM requires reports (including pictures) of the situation at a specific location along the river Sava. Several fire fighters with the right capability type are deployed along the river Sava to monitor the situation. Every fire fighter has an associated proxy that publishes (i.e. advertises) the right service type 'Monitoring Service' producing reports about the current situation. As the crisis manager is only interested in information about the Sava river coming from a specific area (see the oval in Fig. 9.6), not all fire fighters should receive the request. Only fire fighters that are located within the defined area should reply with a report about the situation they observe.

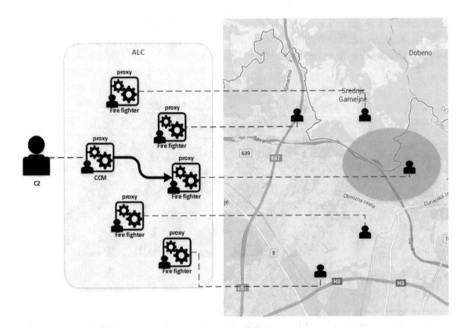

Fig. 9.6 Contextualized request and delivery of information

The provided context of the request is used by the ALC to establish an information flow between the proxy representing the fire fighter in the defined area and the proxy associated with the CCM as shown in Fig. 9.6. Information exchanged using this information flow assumes the geographical context that was provided in the request. Additionally, the crisis manager can also send out a second request for a different area. This will result in the creation of a separate information flow between the CCM proxy and the proxies of the fire fighters that are located in the area specified as the context for the second request. The information exchanged between the proxies for the different requests is completely separated by using dedicated information flows. This means that information from one information flow cannot be used in another information flow. Furthermore, as we will discuss in Sect. 9.4, information between these proxies is encrypted, such that only the two parties involved in the communication can access the data.

9.4 The A-Cloud and Security

There are several notable threats when dealing with information sharing and analysis which can be divided into two categories: data breach threats and threats to the data processing architecture. Furthermore, several types of security properties must be considered, including authentication, authorization (access control), confidentiality, and the integrity of the data, the results of any computations and the system. For example, preserving the confidentiality of the data depends on where the data processing occurs, either on trusted hardware controlled by the owner of the data or on shared hardware. These are not new challenges. Extensive work has been produced in related areas, including foundational work on confidentiality [6], Integrity [7], and authorization [8, 9]. This fundamental work has been applied to the topic of secure distributed computing [10–13].

9.4.1 Security Properties

In order to develop a security architecture for the A-Cloud, first the threats to the system must be identified, with respect to these security attributes. Next, mitigations to these threats should be proposed and implemented. In this section, we briefly identify each of the security attributes in turn and identify threats to the A-Cloud with respect to them. We then suggest some appropriate mitigations.

Confidentiality. From a confidentiality perspective, data is stored in the APC and is accessed by the ALC. Threats to the confidentiality of the data are clear, as the APC should be accessible through a standard RESTful interface. Therefore, data transmitted between ALC proxies, and data stored in the APC by ALC proxies, and information transmitted between ALC proxies, must all be secure against interception. The standard approach to preserving confidentiality is to provide mechanisms to cryptographically encrypt data in the system.

Integrity. Integrity entails ensuring that data either in transit or at rest is not tampered with. In the A-Cloud, data will be in transit in the logic cloud, and at rest in the APC. Typically, controls to ensure integrity involve either cryptographic hashing (or message digesting), using algorithms such as SHA-2 or MD5 or Message Authentication Codes (MAC). Hashing algorithms generate a checksum that (ideally) is unique to the inputted data, where a single bit change to the input results in 50% of the output checksum changing.

Availability. Availability involves ensuring that the system remains usable to the intended users. The advantage to a cloud-based approach is that replication can be used to balance the load to ensure that failures, either accidental or due to an attack, are survivable. In the A-Cloud, the underlying data persistence technologies should be selected to ensure high availability depending on the use case.

Authentication. Authentication entails identifying all the actors using the system, human, automated, and services. Authentication protocols, such as SSL/TLS or Kerberos, operate through the use of authentication tokens. In terms of A-Cloud, the threats to the system include malicious users and services. In order to address these threats, users and services need to be mutually authenticated to ensure that only valid users can use the system and only valid services are presented to those users.

Authorization/Access Control. In complex organizations, there is an acute need for discriminatory access to information and resources. One of the most critical factors when distributing information between partners is controlling when, where, and to whom this information is passed.

As data will be generated by different organizations with different security policies, it is important that there exists a system that allows each organization or user to protect their data explicitly. There are two basic approaches to access control: (1) centralized, and (2) decentralized. Centralized solutions involve building controls that prevent unauthorized users from accessing the system and managing their access in a single central location. Therefore, the system controls access to the entire system and essentially 'owns' all of the data. In terms of the A-Cloud, such a solution is particularly difficult as there are many organizations using the system, and one of their basic requirements is that they want to retain control over their own data. In contrast, a decentralized approach, such as content-based security, involves securing the data directly and allowing organizations to control access to it themselves. In effect, data is encrypted in the ALC and stored in the APC. The APC has no specific access control or authentication. Instead, it is expected that this data is freely accessible and is protected solely by the data encryption. Such decentralized approaches allow each organization to define and enforce access control policies themselves.

Non-repudiation. Non-repudiation is the property of a message where the author cannot deny that they created it. The threat to the A-Cloud system is that a user or service produces some data (at the logic cloud) and then claims that they did not. This threat is alleviated through the use of signatures on each action that cannot be later modified without notice. Therefore, whenever a user or service performs an action, they should validate that action, such as using digital signatures.

9.4.2 Securing the A-Cloud

The A-Cloud should provide a number of basic services. Due to the distributed and decentralized nature of the infrastructure, a content-based security (CBS) approach appears to be the best option. In such a system, the ALC uses a CBS service that provides encrypt and decrypt functionality. This service is provided by the Martello [14–16] component. Figure 9.3 shows how this would logically be placed with the A-Cloud architecture.

9.5 Martello

Martello [15] is an approach to defining and implementing an end-to-end MILS (multiple independent levels of security) solution for data distribution systems. This solution aims at protecting the confidentiality and integrity of data objects for their entire lifetime, regardless of the security of the storage and communication media. The fundamental principle within Martello is that data remains under the control of the data owner at all times. This entails not only deciding on the access rights of other users but also retaining the ability to audit accesses that take place. This is managed through the use of standard cryptographic protocols and tools, where the data owners retain control over the keys to their data. Users of the system must negotiate directly with data owners in order to gain access to the data. While it is impossible to guarantee that unscrupulous (but legitimate) users will not intentionally leak sensitive information, the system keeps audit logs that allow post-facto investigations to identify such leaks and thus allow administrative redress. Martello addresses the above-mentioned limitations of existing solutions in the following ways:

- ensure that at least some critical functionality will remain available (e.g. data exchange between registered participants) even when one or more system components fail;
- support non-disruptive changes to group membership;
- support multiple classification systems by multiple authorities;
- ensure that each information owner maintains control over information released or shared with other authorities; and
- define efficient exclusion and revocation mechanisms.

9.5.1 Interaction Model

Martello is implemented using a publish-subscribe mechanism to share information between different organizations. The basic architecture is shown in Fig. 9.7. As can be seen in this figure, Martello exposes four basic services: *Encrypt, Decrypt, Account Validation,* and *Permission Management.* The first two are the most important: they provide applications with the ability to encrypt and decrypt data that they store elsewhere. The underlying components retrieve cryptographic keys as appropriate, and manage the entire key lifecycle (key creation, retrieval, rotation, and revocation). Applications do not need to consider how the cryptographic mechanisms operate.

The other two services are not integrated within the application—instead they are either integrated within existing authorization and authentication systems (such as LDAP/Active Directory). Alternatively, they are managed directly by the OntoWizard tool described in Sect. 9.3.

A Martello system is made up of one or more domain managers, each one controlling (indirectly) access to data owned by that domain. These domain managers provide the key exchange mechanism for data producers and consumers. The domain manager provides the ability to perform authentication, authorization, and non-repudiation within the system. The function of the domain controller can be determined through how the encryption keys protecting the data are generated. The details of how Martello generates and shares keys are explained in detail in [15].

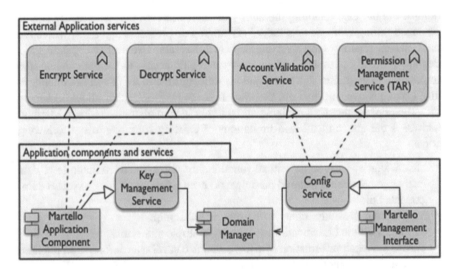

Fig. 9.7 Martello architecture

9.6   Live Testing Exercise

The presented ALC facilitated implementation of the system that supported collaborative analysis at the ATHENA live testing exercise in Slovenia. *The participants from different organizations were provided with a fully functional collaboration and analysis system without the need of any implementation at the exercise location or their organizations.* The experts used the ALC architecture consisting of the proxies shown in Fig. 9.8 that were associated with:

- a C2 operator that receives information from the crisis manager about the developing situation;
- a crisis manager (CCM) that is responsible for the coordination of the overall information collection process in the crisis;
- an engineer from MWC;
- an aquifer expert from NEA; and
- a specially trained fire fighter affiliated to MFD providing professional monitoring service.

These proxies published the capabilities of the experts they represented at the directory service and enabled automated creation of information flows between the right experts on demand. The collaboration was based on automated service discovery; each expert specified the required type of information in a specified area (i.e. context). Such a query was used by the system of DPIF proxies to discover the right experts who could provide the requested information. Whenever a query was generated, all information transmitted to and from the proxies is automatically encrypted using the Martello service. When OntoWizard is used to search for a service, only services that both match the request and the user is *authorized* to access are returned. When a user selects a service, OntoWizard automatically updates the security policy to reflect this choice. This means that there is now a secure virtual information channel between service provider and consumer.

The crisis manager (CCM) specified a request using the browser-based interface shown in Fig. 9.9 that allows to interact with the proxy. The proxy of the CCM sends out a query to the various experts that provide the requested service. In this particular case, the proxies of the MWC, the NEA, and the MFD fire fighter reacted to the specific query related to their service/capabilities. Moreover, the proxies in ALC support targeted querying in the context. For example, the CCM specified an area of interest and queried for the 'MonitoringReport' service in that area (see also the previous, Sect. 9.3.3). This service was provided by the MFD fire fighter, whose operational range included the specified area of interest. This means that the fire fighter was notified via the tablet about the request. Thus, an information flow was automatically established between the proxy of the CCM and the proxy of the fire fighter with the right capability in the range. As soon as the fire fighter reached the specified location, he used the tablet to formulate the report. By clicking the send button the report was automatically delivered to the requester.

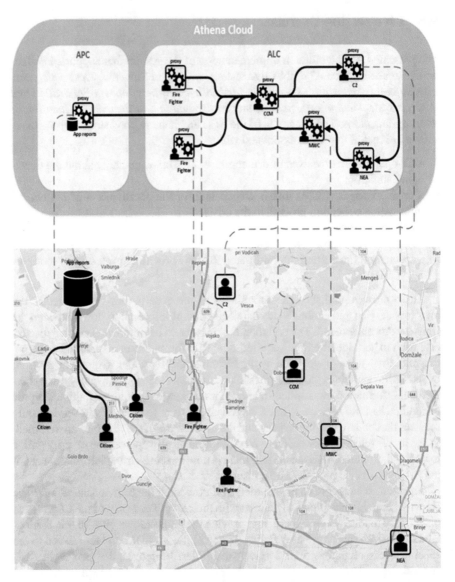

Fig. 9.8 Association of the DPIF proxies to the different stakeholders involved in the ATHENA Ljubljana testing exercise of the ALC about the chemical water pollution incident affecting the aquifer. The *solid directed arrows* in the A-Cloud represent ad-hoc information flows between different proxies of the involved organizations. The dashed lines indicate the user or organization that is associated to the proxy

The DPIF 'App reports' proxy represents automated services that interface the database that collected the reports from the ATHENA app system (part of the APC, as shown in Fig. 9.8). Via this proxy, the reports from the affected area could be filtered and delivered to the requester, in our case, the CCM. For example, reports

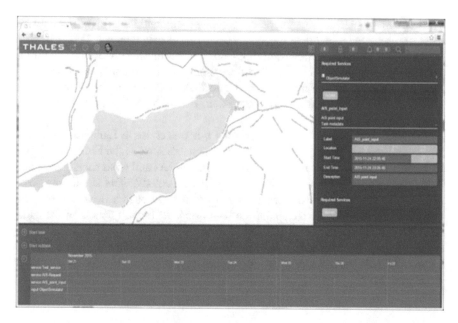

Fig. 9.9 An example of the ALC browser-based expert interface that runs on laptops and tablets

were produced by the members of the fishing association, joggers, and other people along the polluted part of the river using the ATHENA App (see Chap. 7). The App report filtering proxy could extract the reports from the database that were produced in a certain time interval in the specified area.

The DPIF C2 proxy subscribed to analysis reports produced by the CCM. For example, the conclusions/reports from the CCM were delivered to C2 proxy, thus providing the command and control staff with the continuously updated reports based on deep domain knowledge and the latest monitoring reports.

9.7   Conclusion

In this chapter, we have presented the ATHENA Logic Cloud (ALC) approach addressing information management in large-scale crisis situations. The ALC solution supports a cost-efficient, seamless linking of 'deep expertise' provided by humans or automated solutions, and different type of data sources, such as, databases, social media, etc. Such seamless combination of data sources and analysis capabilities is a critical capability required for informed and effective crisis response.

The ALC supports sharing of information under a certain context, such as location, time, clearance, etc. This feature of the ALC provides a unique filtering mechanism that allows to retrieve information that is available or produced under such a context. Contextualized request and delivery of information avoids irrelevant information sharing that could potentially result in loss of valuable time and even incorrect decision making.

The security mechanisms guarantee confidentiality and integrity of data shared between users and allow data owners to decide what data they want to share with others, and for how long they want this sharing to continue. Furthermore, the solution is fully auditable—that is, information owners can determine who has accessed their information. The solution protects information at rest, that is in the Athena Persistent Cloud, and information in motion, that is between the proxies in the Athena Logic Cloud.

The ALC was successfully tested during a live exercise in Ljubljana, Slovenia where different emergency organization participated, such as the Municipal Water Company, National Environmental Agency, fire fighters and crisis managers. Each participating organization was given the task to do collaborative situation assessment in case of a large-scale crisis incident of a toxic chemical release and the potential risk of ground water pollution (see Chap. 12 for full details of the exercise). To facilitate this process of sense making *the participants from different organizations were provided with a fully functional collaboration and analysis system,* i.e. *the ALC, without the need of any implementation at the exercise location or their organizations.*

References

1. Pescaroli, G., & Alexander, D. (2015). A definition of cascading disasters and cascading effects: Going beyond the 'toppling dominos' metaphor. *GRF Davos Planet@Risk, 3*(1), 58–67.
2. Koraeus, M., & Stern, E. (2013). Exploring the crisis management/knowledge management nexus. In B. Akhgar, & S. Yates (Eds.), *Strategic intelligence management* (Chap. 12, pp. 134–149). Butterworth-Heinemann. Retrieved from http://www.sciencedirect.com/science/article/pii/B9780124071919000120.
3. Pavlin, G., Kamermans, M., & Scafes, M. (2010). Dynamic process integration framework: Toward efficient information processing in complex distributed systems. *Informatica, 34,* 477–490.
4. Penders, A., Pavlin, G., & Kamermans, M. (2011). A collaborative approach to construction of complex service oriented systems. *Intelligent Distributed Computing IV, 315,* 55–66.
5. Pavlin, G., de Oude, P., & Penders, A. (2015). *A process integration method and framework.* European Patent WO2015101473. Retrieved from http://worldwide.espacenet.com/publicationDetails/biblio?II=1&ND=3&adjacent=true&locale=en_EP&FT=D&date=20150709&CC=WO&NR=2015101473A1&KC=A1.
6. Bell, D. E., & Padula, L. J. L. (1976). *Secure computer system: Unified exposition and MULTICS interpretation.* Report ESD-TR-75-306. The MITRE Corporation.
7. Biba, K. J. (1976). *Integrity considerations for secure computer systems.* Technical Report MTR-3153 Rev 1. Report ESD-TR-76-372. Bedford, MA: MITRE Corporation.
8. Sandhu, R. S., Coyne, E. J., Feinstein, H. L., & Youman, C. E. (1996). Role-based access control models. *IEEE Computer, 29*(2), 38–47.
9. U. S. Department of Defense. (1983). *Trusted computer system criteria.* Technical Report CSC-STD-001-83, U. S. National Computer Security Center Known as "The Orange Book".
10. Foley, S. N., Mulcahy, B. P., Quillinan, T. B., O'Connor, M., & Morrison, J. P. (2006). Supporting heterogeneous middleware security policies in webcom. *Journal of High Speed Networks: Special Issue on Security Policy Management, 15*(3), 301–313. IOS Press.

11. Foster, I., & Kesselman, C. (1997). Globus: A metacomputing infrastructure toolkit. *The International Journal of Supercomputer Applications and High Performance Computing, 11*(2), 115–128.
12. Kagal, L., Finin, T., & Joshi, A. (2003). A policy based approach to security for the semantic web. In *Proceedings of the 2nd International Semantic Web Conference (ISWC2003)*, Sanibel Island, FL. Springer.
13. White, B. S., Walker, M., Humphrey, M., & Grimshaw, A. S. (2001). LegionFS: A secure and scalable file system supporting cross-domain high-performance applications. In *SC2001: High Performance Networking and Computing*, Denver, CO.
14. Iacob, S. M., Quillinan, T. B., & Van Veelen, J. B. (2015). *A data securing system and method.* European Patent WO2015101474.
15. Pavlin, G., Quillinan, T., Mignet, F., & de Oude, P. (2013). Exploiting intelligence for national security. In B. Akhgar, & S. Yates (Eds.), *Strategic intelligence management* (Chap. 15, pp. 181–198). Butterworth-Heinemann. Retrieved from http://www.sciencedirect.com/science/article/pii/B9780124071919000156.
16. Thales Nederlands B. V. (2015). *Martello: Information security by design.* Retrieved from https://www.thalesgroup.com/en/martello.

Part III
Salient Legal Considerations

Chapter 10
The Relevant Legal Framework

Alison Lyle

10.1   Introduction

The legal framework outlined here addresses data protection and human rights issues that may or will arise when personal information from various sources is collected and processed for use by those managing a crisis. This situation is the one envisaged by the ATHENA project, but is also increasingly relevant in many other situations due to the rapidly increasing use of mobile technology and the overwhelming amount of personal information now available on publicly accessible web-based platforms; in particular, social media sites such as Twitter and Facebook.

Although general issues and principles will be presented, those most pertinent to the use of social media will be focused upon. The context used is the ATHENA proposal of collecting social media data for use by those managing and responding to crises. This will serve to illustrate the key legal requirements that may apply, and which any data controller or processor must comply with.

The legal landscape in terms of data protection at European level is currently an uneven one, which causes uncertainty and inconsistent, inadequate protection. However, the new General Data Protection Regulation became law in April 2016, along with the new Police Directive. Although Member States have 2 years to comply with the new rules, the effect that they will have is also relevant here and will be outlined.

As well as legal requirements, there are other essential factors to be taken account of, which may have an impact upon decisions made and actions taken. These additional considerations include those relating to human rights and citizens' perceptions; these are increasingly important elements in the balance to be achieved between protecting citizens' safety and security, and protecting their rights and freedoms.

A. Lyle (✉)
CENTRIC, Sheffield Hallam University, Sheffield, UK
e-mail: Alison.Lyle@westyorkshire.pnn.police.uk

© Springer International Publishing AG 2017
B. Akhgar et al. (eds.), *Application of Social Media in Crisis Management*,
Transactions on Computational Science and Computational Intelligence,
DOI 10.1007/978-3-319-52419-1_10

10.2   Legal Framework

The proposition on which the ATHENA system rests (see Chap. 5) is the use of data and information from a wide variety of sources in order to create a single picture which will provide accurate and relevant information to those dealing with, and those involved in, a crisis. Without further analysis, it can be said that this action will engage data protection and human rights laws which are fundamental and protected rights at European and Member State levels. In particular, the European Union (EU) and the Council of Europe (CoE) have recognised the renewed importance of these rights and freedoms in the context of the digital age.

The relevant legal instruments, at European level, which will apply to the operation of the ATHENA system, are outlined below.

10.2.1   The Council of Europe

The CoE was established to promote, amongst other things, human rights in the states of Europe and so adopted the European convention on Human Rights (ECHR)[1] which came into force in 1953 and was adopted to further the aim of achieving greater unity between CoE members by safeguarding and promoting human rights and fundamental freedoms. Article 8 of the ECHR protects the privacy of citizens, and will apply to the functioning of the ATHENA system.

In recognition of protecting privacy rights within the context of developing societies, the CoE, in 1981, adopted the Convention for the protection of individuals with regard to the automatic processing of personal data (hereafter referred to as Convention 108).[2] This is still the only international legally binding document in force and has been ratified by all the European Member States as well as the majority of remaining Council of Europe members. In 1999 the EU also became a party.[3] It has influenced all subsequent data protection laws at European level.

Convention 108 is applicable to all data processing carried out by public and private organisations and seeks to protect citizens against violations of their rights in respect of the collection, processing, storage and dissemination of their data. The Convention lays down the requirement for proper legal safeguards in respect of sensitive personal data and provides restrictions on the transborder flow of information. In addition, it provides additional citizen rights to have knowledge about information stored and have access to it if necessary. The principles of fairness, lawfulness

---

[1] Council of Europe, European Convention for the Protection of Human Rights and Fundamental Freedoms, as amended by Protocols Nos 11 and 14, 4 November 1950, CETS 5.

[2] Council of Europe Convention for the Protection of Individuals with regard to Automatic Processing of Personal Data, CETS 108 1981.

[3] Art. 23 (2) of Convention 108 amended allowing the European Communities to accede, adopted by the Committee of Ministers on 15 June 1999.

and proportionality are enshrined in Convention 108 in respect of actions taken by those processing data.

Transborder data flows of personal data were later focused on by the Council of Europe when it adopted an Additional Protocol to Convention 108[4] which recognised the development of exchanges of personal data across national borders and sought to further the protection of citizens in this respect. Additionally, the need for independent regulation of these activities was acknowledged alongside the requirement for Member States to establish independent supervisory authorities to ensure compliance with national laws giving effect to Convention 108 and the Additional Protocol. These independent authorities are given powers of investigation and intervention, to engage in legal proceedings and decide claims regarding the protection of rights associated with data processing.[5]

In recognition of technological advancements and, in particular, the practice of 'profiling', the CoE adopted the Profiling Recommendation[6] which addressed concerns about individuals being placed in predetermined categories without their knowledge by personal data being automatically processed using a variety of software. The CoE aimed to reduce the potential for violations of the rights relating to non-discrimination and dignity enshrined in the ECHR.

10.2.2   The European Union

Within the EU, data protection is a fundamental right enshrined in the Charter of Fundamental Rights of the European Union[7] (The Charter), which was given treaty status by the Treaty of Lisbon[8] in 2009, and the Treaty of the Functioning of the European Union[9] (TFEU). The closely related fundamental right to privacy is also a core principle at this level, protected by the Charter.

The main legislative instrument for data protection within the EU is Directive 95/46/EC[10] (hereafter the Data Protection Directive), which was created with the

---

[4] Council of Europe, Additional Protocol to the Convention for the protection of individuals with regard to automatic processing of personal data, regarding supervisory authorities and transborder data flows, CETS 181 2001.

[5] Ibid., Article 1.

[6] Recommendation of the Committee of Ministers to member states on the protection of individuals with regard to automatic processing of personal data in the context of profiling, CM/Rec (2010)13.

[7] Charter of Fundamental Rights of the European Union [2010] OJ C 83/02.

[8] Treaty of Lisbon amending the Treaty on European Union and the Treaty establishing the European Community, signed at Lisbon, 13 December 2007 [2007] OJ C 306/01.

[9] Consolidated Version of the Treaty on the Functioning of the European Union [2012] OJ C 326/50. This Treaty is the former Treaty establishing the European Community (TEC) which was amended, renamed and renumbered by the Treaty of Lisbon. Article 16 refers to data protection rules.

[10] Directive 95/46/EC of 24 October 1995 on the protection of individuals with regard to the processing of personal data and on the free movement of such data [1995] OJ L 281/31 (Amended by

objective of safeguarding the fundamental rights and freedoms of European citizens in respect of data processing and sought to '…give substance to and amplify'[11] Convention 108.

Directive 2002/58/EC[12] (hereafter the 'E-Privacy Directive') is set out to provide for protection of citizens' privacy in respect of personal data processed in the electronic communications sector. It translates the provisions in the Data Protection Directive and accounts for the development of the electronic communications services and the risks this may pose for the user in respect of their personal data and privacy. It seeks to balance these considerations with those of the continuing promotion and development of electronic communications in the interests of furthering the aims for the internal market. The E-Privacy Directive has been amended by Directive 2009/136/EC,[13] in particular, introducing an obligation in respect of data breach notification.

10.3   Data Protection Principles

The overall picture of rules and principles of data processing can be summarised as follows:

1. It must be *lawful*. Because data processing inherently infringes on human rights, the limitations on this rights set down in s52 of the Charter and s8 of the ECHR necessarily apply. There must be a justified interference with personal rights. The processing must be in *accordance with the law*, have a *legitimate purpose* and be *necessary in a democratic society* in relation to the specific purpose.
2. The *purpose* of the processing must be specifically defined prior to the operation. Any further processing is considered a separate purpose and subject to the same rules and restrictions. Transfer to a third party is a separate purpose.
3. All processing operations are subject to requirements relating to *data quality*. This incorporates the requirements that the data must be *relevant*, *adequate* and *not excessive* in relation to the specific purpose. The principle of *limited retention*

Regulation (EC) of 29 September 2003 1882/2003 adapting to Council Decision 1999/468/EC the provisions relating to committees which assist the Commission in the exercise of its implementing powers laid down in instruments subject to the procedure referred to in Article 251 of the EC Treaty [2003] OJ L 284/1).

[11] Ibid. (11).

[12] Directive 2002/58/EC of 12 July 2002 concerning the processing of personal data and the protection of privacy in the electronic communications sector (Directive on privacy and electronic communications) [2002] OJ L 201/37 (as amended).

[13] Directive 2009/136/EC of 25 November 2009 amending Directive 2002/22/EC on universal service and users' rights relating to electronic communications networks and services, Directive 2002/58/EC concerning the processing of personal data and the protection of privacy in the electronic communications sector and Regulation (EC) No 2006/2004 on cooperation between national authorities responsible for the enforcement of consumer protection laws OJ L 337/11.

applies, stipulating that data must only be kept for the minimum time necessary. Any retained data must be *accurate* and *up to date* in relation to the purpose.

4. Data must be processed *fairly*. The principle of *transparency* is inherent in this requirement. Data subjects must be *informed* of all processing with sufficient detail and clarity. Data subjects have the right to access, update and delete their data.

5. Data controllers must be *accountable* for all their processing operations. This includes taking responsibility for data security, compliance with all data protection requirements and be able to demonstrate this to all stakeholders.

10.3.1 New Legislation

The Treaty of Lisbon along with Article 16 TFEU has created a new legal basis for a more comprehensive and modern approach to data protection, which also covers police and judicial cooperation in criminal matters. This foundation, along with the revelations made by Edward Snowden about surveillance, has led to proposals for new legal instruments at European level and a proposal to update Convention 108 by the CoE.[14]

The EU data protection reform package consists of a General Data Protection Regulation (GDPR)[15] and a Police Directive[16] which were presented in 2012 and became law in April 2016. Member States have until 2018 to comply with the new legislation. Within European law, Regulations are directly applicable whereas Directives are implemented through individual national laws; therefore, the provisions of the GDPR will apply equally across all European States, thereby achieving the much needed harmonisation and legal certainty for all involved.

Within the GDPR there is a degree of continuity in that the developed principles and concepts remain, albeit with some clarification and minor changes. However, there is a stronger implementation of the principles and enforcement of rights and obligations. In particular, the transparency principle and the data minimisation principle are emphasised and there is greater responsibility on the data controller in respect of these. The 'Right to be Forgotten' principle, derived from a ruling by the CJEU,[17] is also included, meaning that any retained data must be deleted as soon as the legitimate purpose for processing ends. In the ATHENA context, this may

---

[14] Overview at: http://www.coe.int/en/web/portal/28-january-data-protection-day-factsheet.

[15] Regulation (EU) 2016/679 of the European Parliament and of the Council of 27 April 2016 on the protection of natural persons with regard to the processing of personal data and on the free movement of such data, and repealing Directive 95/46/EC (General Data Protection Regulation)

[16] Directive (EU) 2016/680 of the European Parliament and of the Council of 27 April 2016 on the protection of natural persons with regard to the processing of personal data by competent authorities for the purposes of the prevention, investigation, detection or prosecution of criminal offences or the execution of criminal penalties, and on the free movement of such data, and repealing council Framework Decision 2008/977/JHA.

[17] Case C-131/12 Google Spain SL, Google Inc. v Agencia Espanola de Proteccion de Datos (AEPD), Mario Costeja Gonzalez [2014].

impact on the possibility of using data for other purposes after the crisis has ended. Citizens' rights in respect of profiling are also strengthened, which also may impact on the ATHENA proposals. A Data Protection Officer must also be appointed where processing is carried out by a public authority.

The CoE reform of Convention 108 began in 2014 and will continue in 2016. The importance of the Snowden revelations and recent Court rulings[18] have reaffirmed the need for the modernisation and strengthening of this internationally binding treaty.[19] The modernisation aims to complement the new EU legislation which brings global harmonisation, in respect of data protection, one step nearer. The principles of proportionality, accountability and transparency are again emphasised and strengthened in the proposals and safeguards for the protection of data subjects are increased.

10.3.2   Exemptions

Arguably, one of the potential legal foundations for the processing of personal data in the ATHENA context rests on one of the exemptions provided for by the legislative instruments. However, it must be remembered that these are strictly applied and will be construed narrowly by the courts. The data protection principles will still be of overriding importance; therefore, each situation must be assessed individually.

The Data Protection Directive contains exemptions under Article 7 which provide for instances where processing is carried out to comply with a legal obligation or to protect the vital interests of the data subject. It is arguable whether the processing of social media data would satisfy this requirement, due to the necessity and proportionality requirements.

Recital 30 of the Data Protection Directive refers to carrying out processing of personal data '…for the performance of a task carried out in the public interest or in the exercise of official authority, or in the legitimate interests of a natural or legal person, provided that the interests or the rights and freedoms of the data subject are not overriding;' and Recital 34 refers to derogation from the prohibition on processing sensitive personal data '…when justified by grounds of important public interest…'[20] The question of how this necessity principle is interpreted and applied was considered by the CJEU in the case of Huber v. Germany,[21] where it was held that a national register which contained personal data of non-nationals for the purposes of fighting crime, did not satisfy the requirement of necessity in the Data Protection Directive, (in light of the TFEU, which prohibits discrimination based on nationality).[22]

---

[18] For example; Case C-362/14 Maximillian Schrems v Data Protection Commissioner.

[19] An overview is presented at: http://www.coe.int/en/web/portal/28-january-data-protection-day-factsheet.

[20] See also Article 7(e) of the Data Protection Directive.

[21] Case C-524/05 Huber v. Germany [2008] ECR I-9705.

[22] TFEU, Article 18.

Article 52 of the Charter limits the exercise of rights and freedoms but '…only if they are necessary and genuinely meet objectives of general interest recognised by the Union or the need to protect the rights and freedoms of others'.[23] However, if the aim can be achieved in a less intrusive and more proportionate way, then this must be done.

The legitimate interests of the controller would also allow for the processing of personal data without consent.[24] Whilst Article 7 of the Data Protection Directive sets out an exhaustive list of principles, the degree to which the individual's rights are infringed can depend on whether the data processed already appears in public sources. This is significant in relation to the balancing act that must be carried out between the legitimate interests of the controller and those of the data subject. This may apply to the data processed by the ATHENA system; if it could be said that posting information on a social media site could be deemed 'public' this would be balanced against the legitimate interests of the ATHENA controller.

In the new GDPR the exceptions in Article 6 specify the criteria for lawful processing, address the balancing of interests and compliance with legal obligations and public interest. Article 7(f), which provides for legitimate interests of the controller, refers to particular consideration for children when considering balance of interests. This new provision also specifically states that it will not apply to public authorities in the performance of their tasks; this may affect the ATHENA operation if the data controller is a public authority.

10.3.3   Member States' Laws

The Data Protection Directive was designed to allow flexibility to Member States in their implementation of it. However, this has also produced a lack of uniformity which in turn creates uncertainty and uneven protection. In the so-called 'digital age' these differences have more of an impact as traditional and geographic boundaries become less significant and citizens' lives are increasingly lived in the borderless cyber-environment. When fully implemented, the new GDPR will go a long way in creating legal certainty and harmony, but in the meantime data controllers will need to be aware of restrictions and requirements in different countries.

10.3.4   Jurisprudence

The effectiveness of laws is determined by the way in which they are enforced. In the uncertain and uneven data protection landscape, the jurisprudence of the European courts serves to clarify concepts and create a degree of consistency.

---

[23] The Charter, Article 52(1).

[24] Article 7(f) of the Data Protection Directive.

The European Court of Human Rights (hereafter ECtHR) was set up in Strasbourg in 1959 to ensure that parties to the ECHR fulfilled their obligations. The Court hears cases from individuals, groups, Non-Governmental Organisations (NGOs) or legal persons alleging breaches of human rights under the Convention. The issue of data protection has arisen and been dealt with many times and has established that in addition to acting in accordance with the ECHR, States also have an obligation to take positive action to ensure these rights are protected. The Court employs various methods of interpretation when deciding cases[25] but generally takes a teleological approach in which the object and purpose of the ECHR inform the decision. The evolutive approach also plays a part however, where the ECtHR will take into consideration changing conditions within the contracting party States. These combined approaches ensure that the ECHR retains its purpose in developing society.

The Court of Justice of the European Union (hereafter CJEU) was established to ensure Member State compliance with European legal instruments. It will decide whether national laws concerning data protection issues are valid and correctly interpret the meaning of the Data Protection Directive. It is an important point that data protection is a right which will be balanced against other considerations; it is not absolute. In joined cases Volker und Markus Schecke GbR and Hartmt Eifert v. Land Hessen[26] at paragraph 48 of the judgment, the CJEU reiterated that the right to data protection has to be considered within the context of its function in society.

10.4   ATHENA

Four of the main components of the system are briefly explained here, so that some of the potential legal issues in relation to each can be identified.

10.4.1   The Crisis Mobile App

The ATHENA app involves an ongoing, mutually beneficial exchange of information between those managing the crisis, citizens and first responders 'on the ground'. The information provided by the user would be given with their consent and would involve sending photographs and videos of the scene and providing textual information about the unfolding events. This may include images of trapped, injured or deceased persons, and video and sound recording of others in the vicinity. All these are capable of identifying individuals and so would be defined as personal data, some of which would be sensitive. The protection rights would relate to the subject

---

[25] As discussed in H Senden Interpretaion of Fundamental Rights in a Multilevel Legal System (School of Human Rights Series, Vol 46, Intersentia, 2011).

[26] Joined cases C-92/09 and C-93/09 Volker und Markus Schecke GbR and Hartmt Eifert v. Land Hessen [2010] ECR 000.

of the information provided, rather than the provider. Geo-spacial and temporal information would be included and should be with the user's consent (see Chap. 7).

Information received via the app would be simplified and concise, relating to such things as danger zones, safe routes and the seriousness of the incident. Users would receive information appropriate to their user tier and the classification of the information. Potential issues arising here relate to the use of a ranking system based on pre-authorised and learned credibility ratings which could be described as 'profiling' and is a significant area of concern in relation to data protection.

10.4.2   Social Media

As well as collecting information from the app, the ATHENA system will use a web crawling tool to identify and collect information from public social media sites, and other open web-based information sources, such as Twitter, RSS feeds and news sites. Dedicated social media sites will also be used by the Crisis Command and Control Intelligence Dashboard (CCCID) operator to post information to citizens.

In the proposed ATHENA system, most of the data will be collected from various social media sites. These sites are provided by different service providers, all of whom have privacy policies in place which the user has agreed to. These policies should stipulate any use to which their data will be put. If ATHENA subsequently collects and processes personal data from these sites, this constitutes a different purpose and raises various data protection issues, such as legitimate purpose, necessity, proportionality and data quality.

The collected data will be subjected to processes which will allocate certain classifications to information according to its type and source and a clearance/credibility level will determine its further handling and dissemination. This may involve analysing personal details attached to a person's profile. Previous social media posts by the data source may form part of this analysis, as well as role or position, religious or political affiliations and indications of intelligence or age. Analysing, judging and acting on this information would raise several issues that may include profiling and discrimination (see Chap. 6).

10.4.3   The Crisis Information Processing Centre (CIPC)

This is a collection of tools which will acquire data from various sources, filter and categorise it so that relevant information can be meaningfully and appropriately used by those managing the crisis. Initially, the filter system will remove irrelevant, and unwanted information using Natural Language Processing (NLP) and the crisis taxonomy system will identify relevant information. Aggregation and analysis tools will carry out tasks such as credibility scoring, Formal Concept Analysis (FCA) summarising, data fusion, classification and clearance identification and sentiment analysis.

ATHENA proposes large-scale data collection and although this process uses filters such as hash tag syntax and crisis taxonomies. These methods would not identify sensitive personal data or vulnerable persons' data, which would inevitably be collected in the process and may raise issues connected with privacy and handling sensitive data. Although irrelevant or unwanted information will be discarded, from the citizen's point of view their data has nevertheless been collected and it will not be known by them whether it has been used or not.

It is proposed that a semantic driven tool and information validation guidelines will identify fraudulent, malicious or hoax information, which may involve using personal profile information. It can be assumed that those who post hoax information do not want their personal details known, therefore it would arguably follow that they would not consent to this processing. This also raises potential human rights issues in respect of persons identified as unreliable who may not be.

10.4.4 Crisis Command and Control Intelligence Dashboard

The information that has been collected, filtered and processes is sent to and disseminated from the CCCID so that those managing the crisis can do so effectively by communicating with all those involved. The roles within the CCCID will involve decision-making about how information is to be used, this would define CCCID operators as 'data controllers' within the legislation and so impose a number of duties and obligations.

The CCCID operators will have a default 'all information' level of clearance in respect of data collected and processed. This feature adds a human element to an otherwise largely automated system, which raises issues. Careful consideration must be given to who will use the CCCID due to unrestricted access to personal and sensitive information. Subjective decision-making in relation to credibility of sources and recipients again raises 'profiling' issues.

10.5 Potential Issues

The issues and concepts that arise out of the analysis of the proposed actions of the ATHENA system that require legal consideration are many; some of them can be summarised as follows:

1. *Personal data*—what is considered to be personal data will determine what legal restraints will apply. Personal data is defined as any information which allows a natural person to be identified either directly or indirectly. Information collected from social media sites would include the username and potentially more information embedded in the user profile, which could enable identification of an individual. Well-known pseudonyms would arguably be as revealing as a real name, and dates of birth are typically included to alert other users to birthdays.

For the purposes of the Data Protection Directive, sound, image and video data in the ATHENA system would be regarded as personal data. However, it would be the personal data of the subject of the recording rather than the person who supplied it, and would thus be collected without the subject's consent. This would arguably be acceptable under the exemption relating to public security and preserving life. However, exemptions will be construed narrowly and any further processing, storage or dissemination outside these strict requirements could not have the same legal basis.

2. *Sensitive data*—under data protection law, certain categories of data require different handling and are subject to separate legal requirements. Sensitive data is identified as such due to the additional risks to the data subjects when processing this information. All legal instruments refer to this special category and put additional obligations on controllers and processors when dealing with this information. Both the Data Protection Directive and Convention 108 refer to categories of sensitive data as that which reveals racial or ethnic origin, religious, political or other beliefs and information relating to health or sexual life.

Data collected on a large scale from social media sites may have sensitive information within it and the processing of this information is prohibited under Article 8 of the Data Protection Directive. However, Article 8 allows for the processing of this information where '...*processing is necessary to protect the vital interests of the data subject or of another person where the data subject is physically or legally incapable of giving his consent*'...[27] which would apply to the ATHENA system where data is collected during a crisis situation in order to provide information to ensure citizens' safety. Another exemption under Article 8 is where '...*the processing relates to data which are manifestly made public by the data subject*...', it might be argued that information from a social media site was 'public'.

3. *Controller and Processor*—the identification of data controller and data processor is crucial in determining responsibility for compliance with legal obligations. Article 2 of the Data Protection Directive defines controller as '...*the natural or legal person, public authority, agency or any other body which alone or jointly with others determines the purposes and means of the processing of personal data;...*'.[28] Processor is defined as '...*a natural or legal person, public authority, agency or any other body which processes personal data on behalf of the controller*[29];' Under Convention 108 the concept of controller also includes one who makes decision on which categories of personal data would be stored.[30] Clearly these definitions would apply to those using the ATHENA system. Under the new GDPR the duties and responsibilities of the data controller are increased and emphasised. In the ATHENA context, where there may be several joint control-

[27] Data Protection Directive, Article 8(c).
[28] ibid., Article 2(d).
[29] Ibid., Article 2(e).
[30] Convention 108, Article 2(d).

lers, it might be preferable for the company or body to be considered the controller with individuals acting under its direct control.

4. *Processing*—establishing the type and purpose of processing is fundamental to carrying out lawful operations. There must be a legal basis for any collection and processing of personal data; within the ATHENA context the alternatives may be consent or one of the exemptions provided for in the legislation. The exceptions provided under Article 7 of the Data Protection Directive may not be the appropriate legal basis for legitimate processing of large amounts of data collected from social media networks. Such large-scale interference with individuals' privacy would arguably not satisfy the requirement relating to principles of necessity and proportionality under this provision. An additional consideration would be that when carrying out the required balancing act between legitimate interests of the controller and the individuals, those of the individuals may outweigh those of the controller. It could be argued that the legitimate aim of the public authority or controller can be achieved through less intrusive means. An alternative legal basis for this processing would be consent. It may be that the use of hashtags would satisfy the requirement of implied consent.

5. *Dissemination*—this is considered to be one of the elements of processing and is subject to the same legal constraints. The status of 'recipient' and 'third party' needs to be established to ensure compliance with data protection law. In respect of the re-use of processed information by way of posting it on a dedicated website, it is important to note that the information is still considered to be personal data if the data subject is identified or identifiable and therefore is still subject to data protection rules. This use of the data would need to be specified as part of the original purpose and satisfy the data protection principles. It is likely that the test of whether it could reasonably be understood and foreseen as part of the original purpose for processing might be satisfied if the website was for providing information to the public in order to protect their safety. If the compatibility test were not satisfied, then Article 13 of the Data Protection Directive allows for '*…a necessary measure to safeguard…public security..*' which may apply.

6. *Profiling*—this type of processing raises particular concerns. The credibility and ranking system forms part of the ATHENA data collection and processing operations by applying predetermined criteria to information obtained from data sources, and could have the effect of rejecting or disregarding potentially valuable information from data sources based largely on their writing skills, apparent age and other information embedded in user profiles on social networking sites. This has the potential for serious and unjustified discrimination against individuals who not only may be denied the opportunity to contribute potentially valuable information during a crisis, but who may also be unfairly categorised as 'unreliable'. The result of this profiling may not be communicated to the individual and so may violate their fundamental right of non-discrimination enshrined in the Charter and in the ECHR.

10.6   Additional Considerations

Due to the legal status of both the ECHR and the Charter, all Member States have an overriding obligation to consider and protect fundamental rights and freedoms. It might be the case that data which does not fall into the category of personal data in respect of the Data Protection Directive, and its implementing national laws, may nevertheless engage the ECHR in respect of the right to privacy, freedom to impart and receive information, freedom of expression and the right to non-discrimination. In addition, the Charter may also apply to some situations where the data protection rules would not apply.

The jurisprudence behind these laws serves to protect, uphold and achieve freedoms and fairness. However, fairness and freedom are essentially subjective concepts without universal definition. What one individual or group perceives as fair can differ enormously from another. The Snowden revelations show how quickly public reaction can escalate and instigate positive action in respect of human rights; various action groups continually campaign to raise awareness among citizens and actively promote and protect their rights. Due to the ATHENA operation depending wholly on the participation of the public, it is important to consider citizens' perspectives in relation to relevant issues.

One action group is 'Big Brother Watch', who describe themselves as a 'campaign group for the digital age'[31] and was set up in 2009 to challenge policies perceived to threaten privacy, freedoms and civil liberties. At the time of writing, this group (along with additional applicants) has a case before the ECtHR which addresses a concern raised by the Snowden leaks in the press, namely that they have been the subject of surveillance by GCHQ which is not in accordance with the law and interferes with privacy rights under Article 8 ECHR. A further contention is that generic interception of communications via fibre-optic cables is a disproportionate interference with the private lives of potentially millions of people. The case, although designated a priority, has been ongoing since 2013 and after an adjournment is now moving ahead. The UK Government has been asked to respond. An indication of the strength of feeling about these issues is illustrated by information posted on the Big Brother Watch website[32] which states that over 1400 people donated £27,279 in 48 h to support the case.

Another case which attracted large-scale public response involved Max Schrems, who took action against Facebook over the transfer of personal data to the US.[33] The case made significant history in 2015, when the CJEU declared the Safe Harbour[34] agreement invalid and that data protection authorities could suspend data transfers to third countries if they violated EU rights. This has led to changes in the law at European level in respect of transferring personal data. The scheme that replaced

---

[31] 'About us' page at: www.bigbrotherwatch.org.uk/about.

[32] https://www.privacynotprism.org.uk/.

[33] Case C-362/14 Maximillian Schrems v Data Protection Commissioner.

[34] A regime regulating data flows from the EU to the US.

the invalidated 'Safe Harbour' agreement, the EU-US 'Privacy Shield', is now being challenged by privacy action group Digital Rights Ireland, who allege its inadequacy in respect of protecting individuals' privacy rights.

These are just two of many examples which illustrate the strength of feeling by many citizens in respect of their fundamental rights, and perceived surveillance by the State. This makes it all the more important for ATHENA, which may be operated by law enforcement authorities, to carefully consider not only the legality of their actions, but the way they might be perceived by the public.

10.7 Conclusion

The issues outlined and briefly addressed here apply within the ATHENA context, but also on a more general level and particularly in respect of personal data from social media sources. Social media sites are recognised ways of facilitating social cohesion through mass communication, and allow many individuals to share and have access to information. It is on this understanding that the ATHENA project is based. These sites also promote the enjoyment of the fundamental rights and freedoms which are the cornerstone of European law, such as the freedom of expression and to receive and impart information. Just as the Internet has the power to allow and promote these rights and freedoms, it follows that there is the potential to damage and restrict them by the same means. The rules that exist, and particularly the ones for the future, are designed not as barriers but as a mechanism to achieve a fair balance between legitimate rights.

It can be seen that data protection and the associated human rights and fundamental freedoms are at the core of European concerns and considerations. The overriding principles, consistently applied by both the CJEU and the ECtHR, will be emphasised in the GDPR and clearly illustrate the importance that is attached to protecting the interests of citizens in the digital age.

It is crucial that ATHENA is recognised not only for protecting the lives of citizens during a crisis, but protecting their personal data; it should be seen as not only safeguarding people's security but safeguarding their rights and freedoms too. The ATHENA system relies upon the support of citizens for its successful operation and in turn those citizens should be able to rely on ATHENA's integrity and respect.

Chapter 11
Legal Considerations Relating to the Police Use of Social Media

Fraser Sampson and Alison Lyle

11.1 Introduction

The impact of social media on emergency management has been substantial [1] and its 'growing ubiquity, not only in geopolitical, economic and business spheres but also in official responsiveness to crisis and disaster', has been well documented [2]. As preceding chapters have discussed, the ATHENA project will develop a system to allow the public to play a part in the effective and efficient management of a crisis situation by contributing to the 'conversation' through their use of social media networks and hi-tech mobile devices. Enabling the public to have a voice in such situations is a valuable asset, not only to those managing and responding to the crisis, but also by empowering communities to help themselves and communicate their needs (see Chaps. 5, 12 and 13). However, there are also opportunities for police to use the data collected by the ATHENA system to enable more effective and efficient investigations into criminal offences that occur at the time, or as a consequence, of the crisis. Additionally, there is a proposition that data collected throughout the crisis will be retained and shared with European LEAs. It can be seen then that there are several parties to whom the ATHENA system will be of benefit, and this theme of mutual benefit will run throughout this chapter, which addresses the legal issues that may or will arise through police use of social media, both within and related to the ATHENA context.

The concept of privacy has evolved steadily alongside changing practices of the societies in which we live, but with the rise in numbers of people living out their

F. Sampson
Office of the Police and Crime Commissioner for West Yorkshire, UK

A. Lyle (✉)
CENTRIC, Sheffield Hallam University, Sheffield, UK
e-mail: Alison.Lyle@westyorkshire.pnn.police.uk

© Springer International Publishing AG 2017
B. Akhgar et al. (eds.), *Application of Social Media in Crisis Management*,
Transactions on Computational Science and Computational Intelligence,
DOI 10.1007/978-3-319-52419-1_11

lives in 'cyberspace', in the virtual world of social media, it has taken on new meaning for many people. Social media users typically post information, photographs and videos that reveal personal and often sensitive aspects of their lives, yet as recent surveys show [3], these same people are very concerned about who sees this information and, more importantly, who collects and uses it. Although social media platforms are perceived as public, free-for-all environments by organisations wishing to make use of the data, those who post information expect their privacy to be respected. This difference in perceptions is a crucial consideration for any law enforcement agency if they are to successfully share the social media world with the general population and make use of the valuable resource that it is. The key element to be established is trust; any successful relationship is built upon this and the one between the police and citizens is no different in this respect. Overall, police use of social media is seen as positive and of benefit to all those involved. However, the risk of certain activities being seen as surveillance by the State is high; this is a very sensitive area in the public arena.

ATHENA expressly recognises the existence and importance of legal considerations relating to privacy and data protection [4]; what follows is intended to help identify the legal considerations and assist in the research that will flow from the project. By contributing to the requisite legal—and ethical—framework required within ATHENA, this chapter also aims to provide a useful platform for discussion of any future initiatives of this type.

11.2   ATHENA

ATHENA is exploring the ways and extent to which LEAs and other crisis responders might possibly harness new communication media—particularly web-based social media such as Twitter and Facebook—to provide efficient and effective communication and enhanced situational awareness during a crisis (see Chap. 5).

The various elements of social media constituting the ATHENA approach to aiding citizens and LEAs in response to crises may usefully be illustrated by analogy using a process from chemistry rather than law. Stoichiometry is an activity (or exercise) involving the close analysis of the different relationships between relative quantities of elements taking part in a chemical reaction and is almost a perfect metaphor for the legal issues arising in ATHENA. The social reactions caused by 'crises' of the type envisaged by ATHENA (particularly where there is an investigative or criminal justice element) create legal relationships between the many parties and agencies involved and require an appropriate legal equation balancing; on the one hand, preserving/ensuring the safety and security of those involved and on the other, the observance of fundamental rights, including privacy, and legal compliance around data protection.

The reactive communications of citizens caught in crises already go way beyond the passive, information-consuming audience that the police [4] and press have previously been used to encountering. ATHENA seeks to invite those same citizens into providing and accessing a network of potentially limitless operational data. This is one of its innovations and, at the same time, a significant legal consideration.

ATHENA will use social media during a crisis in two ways:

First, it will be used by one of the CCCID operators as a tool to send information out to citizens. In this way, those managing the crisis will be able to reach the maximum number of people as quickly as possible. ATHENA has two dedicated social media pages for this purpose, one on Twitter and one on Facebook. Each one is directly accessible from the ATHENA app (see Chap. 7).

Secondly, ATHENA will use web-crawling software to collect information from social media platforms[1] using algorithms, Natural Language Processing (NLP) and hashtag syntax. This will automatically identify social media posts that are relevant to the crisis. These will then be further processed, analysed and assigned credibility and priority ratings and fed into the CCCID as either aggregated reports or individually. Social media posts will be from the general public and those citizens acting as pre-first responders in the crisis, accessing social media pages through the buttons on the ATHENA app.

Moreover, ATHENA is planning to go much further. For example, not only it is planning to analyse geotags to show where individuals were at the relevant time(s), or word collocation (where the frequency of occurrence of pairs or groups of words occurring in proximity is determined [5]); the team is going to 'move the semantic analysis of social media data beyond current state of the art'. The project will use automated processes to conduct sentiment analyses to locate and analyse digital content in real time to determine the contributor's 'emotional meaning', developing 'credibility assessments' and 'scoring tools' to underpin the use of ATHENA data mining, social network and sentiment analysis tools to tag messages with reliability scores.[2] Such sensitive and intrusive processing will, it is submitted, require an elevated degree of trust between data controller and data subject, and the latter will need to be in no doubt what level of analysis of their data has been signed up to. In that regard the ATHENA team will require clarity and a legal basis for the action. In taking part in the 'reaction' to civil contingency and accepting the 'invitation' to participate, citizens as elements will find themselves in legal relationships—both direct and vicarious—which are likely to prove operationally hazardous. ATHENA is therefore under a profound obligation to ensure that such relationships, and their possible implications and consequences, are adequately considered and catered for.

Any consideration of the legal relationships arising from ATHENA, and in particular police use of social media, needs to begin with the European regulatory framework governing human rights, data protection and the use of personal data in the police sector. The general protection, processing, sharing and retention of data are heavily regulated by both European and Member State law, some of which creates particular challenges and dilemmas for law enforcement authorities (LEAs) [6]. However, before the key issues that need to be included in this discussion are identified, it is useful to consider the various ways in which the police use social media and the growing importance of this new policing area.

---

[1] Facebook and Twitter.

[2] see ATHENA submission Sentiment and reliability analysis—B.1.1.1.

11.3   Police Use of Social Media

The increasing use of social media has created a new, virtual environment for individuals and communities to live out part of their lives and has thus changed the way policing needs to be carried out. In order to keep the public safe and to detect, investigate and prevent crime the police need to be present in the new public space, in the same way as a geographical space, to ensure the safety of the public, provide reassurance and prevent or deal with crime and disorder (see Chaps. 2, 4 and 6).

The Police Foundation[3] divides the police service's use of social media into three broad areas [7] as follows:

1. Providing Information—enabling specifically targeted information to be shared quickly, easily and cheaply;
2. Engagement—providing the police with a way of connecting and building relationships with local communities and members of the public;
3. Intelligence and Investigation—allowing the police to listen to what their communities are saying and to build evidence for investigations by monitoring social media content.

In respect of intelligence and investigation work, the police frequently post requests for specific information or post photographs or videos to gain assistance with identifications. Used in this way, social media is a valuable tool for police and an empowering tool for the general public who are directly supporting policing efforts. Police will also use social media sites to gain crucial information which may otherwise have been resource intensive or been unobtainable.[4]

A study from the COMPOSITE project [8] reveals that first results from a study of European police use of social media indicate high levels of general acceptance and perceived usefulness across all forces who responded. The three factors contributing to acceptance were as follows:

1. Usefulness for me as a police officer (improve performance).
2. Usefulness for my police force (improve force performance).
3. Fit with my task (is compatible with all aspects of my work).

The areas of policing ranking these more highly were community policing and crime investigations.

Another survey [9] reported that, on a global level,[5] citizens have the same view and are enthusiastic about being involved with the police in a digital way. Overall, 79% of respondents were in favour of more digital interaction with the police and 72% were more willing, than a year before, to engage with the police using social

---

[3] The Police Foundation is the only independent charity that acts as a bridge between the public, the police and the Government in the UK, while being owned by none of them. See: www.police-foundation.org.uk.

[4] See the COMPOSITE project at: http://www.composite-project.eu/.

[5] Countries surveyed were Australia, France, Germany, Singapore, Netherlands, Spain, UK and the United States (Accenture 2014, p. 9).

media. It must be pointed out, however, that interaction on a voluntary basis is very different from large-scale collection of data from these sources, without citizens' knowledge or consent. This crucial difference forms the fine line between positive cooperation and perceived (or actual) surveillance.

Much of the relevant data that will be collected, processed and retained by ATHENA emanates from social media. There has been some significant research in the realm of social media and the policing of disorder generally, mostly focusing on the role played by communication in the mobilisation of disorder and coordination of participants [10].[6] There is also formal guidance for the police [11] though this gives no specific advice on personal data protection and compliance of the type being described here.

The relationship between public disorder and the State use of social media has largely developed around the possibility of governments using their powers to censor or curtail communication as a means of suppression [12]. However, the advances being offered by ATHENA intend to shift this and make the use of social media by crisis responders a central tactical and strategic plank.

ATHENA considers South Yorkshire Police's social media strategy during protests around the Liberal Democrats' 2011 Spring Conference in Sheffield [13] and the force's use of Twitter and Facebook to interact with members of the public. ATHENA contrasts the police use of social media during the TUC's 'March for the Alternative' in 2011, which received praise from independent observers [14] but also criticism on the basis that the police seemed more concerned with managing public perception than facilitating communication, causing mistrust and unrest [15]. They also compare the strategies of the Metropolitan Police Service (MPS) and Greater Manchester Police (GMP) during the riots of 2011, highlighting the 'relative success' of GMP's more 'expressive' approach compared to that of the MPS's 'instrumental' strategy [16]. ATHENA tracks how British police forces not only saw a tremendous growth in the number of Twitter followers but how they also, for the first time, engaged with the public on a large scale via social media, using Twitter as the main platform' [17].

ATHENA notes the work of researchers who found that the MPS's use of social media was hampered by the lack of a coherent social media strategy and of appropriate resources, failing to take advantage of the increasing number of people who—as events unfolded—followed the police on Twitter, creating a growing capacity for communicating risk and communicating about risk [18]. These authors observed varied approaches between the two police forces, citing one Metropolitan Police officer who concluded they had not been 'wholly up to speed in using social media as an intelligence tool, an investigative tool and most importantly as an engagement tool' [18, p. 21]. By contrast, Greater Manchester Police were congratulated on the way that they had chosen to use such media during the riots [18].

While Manchester was less affected by the riots, their local police force had already established a reputation for embracing Twitter and had experimented with

---

[6] For example, the French and Grecian riots of 2005 and 2008 as well as the riots and social disorder around the so-called 'Arab Spring' [10].

its use in campaigns before, and ATHENA contrasts how the two forces had made use of social media during the disturbances. For example, during the period from 4 to 13 August, the MPS posted 132 tweets, but GMP almost three times as many (a total of 371). Beyond the quantitative difference, there was also significant qualitative variation within the content and style of messages. The MPS's clear preference for using a much more impersonal style, directed to a generic audience as opposed to individual followers. While both forces employed Twitter primarily to gather and disseminate information about the riots (e.g. by posting CCTV images of perpetrators on Flickr and leaving phone numbers and web site addresses), GMP placed a much greater emphasis on reassuring the public—i.e. 'noting that everything was calm and the public should not worry'. The MPS, by contrast, focused principally on maintaining law and order illustrating how one force's approach followed an instrumental strategy while the other's was primarily expressive.

While undoubtedly apprehending an innovative and sophisticated approach to social media by LEAs, the plans by ATHENA to exceed anything that has previously been done with the data generated by crisis relationships will need to be drawn to the attention of any crisis responder; arguably, it needs to be publicised to communities at large. In other words, they too will be well advised to adopt an expressive strategy before drawing upon the vast social media capacity and taking on the reactivity of digital live-time communications. The parameters of their 'compatible purposes' will be particularly important in the setting of public disorder where protracted investigative process usually follows the settling of the dust. Criminal investigations can continue for months or even years beyond the emergency itself and the availability of relevant responders' data to be analysed, retained and shared with other agencies raises substantial legal issues, particularly where that data might be used for purposes that are adverse to the responder's individual interests. It is recommended that ATHENA makes explicit any likelihood that responders' data may be used for criminal intelligence, investigation and even prosecution purposes. Further legal issues arise in the case of political protest where LEAs are often as interested in upstream prevention as they are in real time responding.[7]

The very transient nature of digital relationships that coalesce around an event such as a public disturbance can be seen from analyses of social media patters such as Twitter (see Fig. 11.1), whereby once the event/activity/interest that unites members diminishes, so does the digital 'community' itself [19]. In light of this, it is proposed that ATHENA takes account of and prepares for the retention/deletion of relevant data once the uniting crisis (or at least an agreed phase of it) has passed.

Where disorder emanates from political protest, the challenges for LEAs increase—and so they will for ATHENA. ATHENA has analysed a number of scenarios including the use of social media to create 'flash mobs' with thousands directed to riot hotspots such as Millbank [20]. The volume of social media

---

[7] R (on the application of Laporte) v Chief Constable of Gloucestershire [2006] UKHL 55; where it was held that the police action to prevent the applicant travelling from Gloucestershire to an anti-war rally in London interfered disproportionately and therefore unlawfully with the applicant's Convention rights of freedom of expression and assembly.

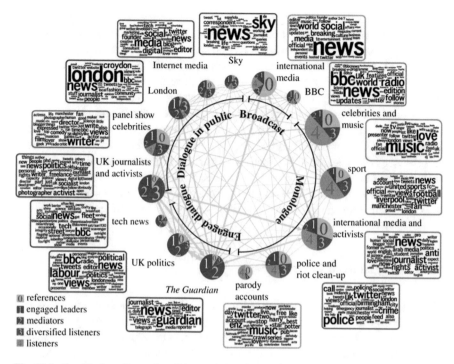

Fig. 11.1 Graphic from Beguerisse-Díaz et al. (op cit.) conflating 15 interest communities of the 'most influential Twitter users' during the 2011 riots in London

interaction and the generation of attendant relationships from another episode of disorder in London disorder illustrate the size of the task, with some 2.6 million Tweets posted from approximately 700,000 distinct user accounts between 1 pm on 6 August and 8 pm on 17 August 2011 [21].

In these settings the tactics of the police have produced a series of legal challenges, and demonstrated how difficult it is to achieve the fine balance between the obligations of the State to ensure the security and safety of its citizens and its duty to ensure the protection of their human rights and fundamental freedoms[8]. Even where the relevant event takes place in public, the recording and retention of personal data about individuals involved can nevertheless amount to an unlawful interference with the right to respect for private life under Article 8 of the European Convention of Human Rights. The importance of retaining/regaining the holy trinity of trust, confidence and legitimacy for LEAs, and their citizens is nowhere clearer perhaps than where the public disorder has a political complexion and covert tactics have been deployed.[9]

---

[8] See e.g. R (on the application of Catt) v The Association of Chief Police Officers of England, Wales and Northern Ireland and The Commissioner of Police for the Metropolis [2013] EWCA Civ 192.

[9] See: http://www.thetimes.co.uk/tto/news/uk/crime/article3306515.ece; See also: https://netpol.org/2014/05/19/netpol-ico-complaint/.

As illustrated, there are many ways in which the use of social media by the police is a vitally important means of communication, as well as a rich source of intelligence and information for investigations into specific crimes. The word 'specific' in this context is a key one; processing personal data in relation to specific crimes will likely satisfy the legal requirements. The difference is that ATHENA proposes to carry out web crawling for one purpose (information in a crisis) and to then retain this information and use it for policing purposes; herein lies the problem. This and related issues will need careful consideration if a balance is to be achieved and social media data is both protected and useful for policing purposes. The way to achieve this balance is to be aware of the legislative instruments that are effective in this area, to understand the principles behind those constraints and to adhere to best practices that take all these into consideration. The laws and legal instruments applying to this area stem from human rights laws at international and European level.

11.4   Legislative Instruments

The two sources of data protection at European level, the Council of Europe (CoE) and the European Union (EU), are separate legal systems but are closely related, and each has enacted specific legal instruments that aim to achieve a balance between the competing interests of protecting individuals' data and ensuring national and public safety.

The importance of providing protection for individuals against intrusion by the State was recognised in the United Nations Universal Declaration of Human Rights in 1948.[10] This influenced the CoE European Convention on Human Rights (ECHR) in 1950, Article 8 of which provides that public authorities shall not interfere with the right to private life, except in specific circumstances, which are set out in Article 8(2).[11]

The CoE recognised that more specific protection was needed in the digital age and in 1981 Convention 108[12] was opened for signature. Convention 108 remains the only legally binding international instrument in the area of data protection and specifically covers processing of information by police. Another CoE legal instrument that covers all areas of police work is the Police Recommendation.[13] This

---

[10] Article 12 – 'No one shall be subjected to arbitrary interference with his privacy, family, home or correspondence…..'

[11] Article 8(2)—'There shall be no interference by a public authority with the exercise of this right except such as is in accordance with the law and is necessary in a democratic society in the interests of national security, public safety or the economic wellbeing of the country, for the prevention of disorder or crime, for the protection of health or morals, or for the protection of the rights and freedoms of others'.

[12] CoE, Convention for the Protection of Individuals with regard to Automatic Processing of Personal Data. Council of Europe, CETS No. 108, 1981.

[13] CoE, Committee of Ministers (1987) Recommendation No. R(87)15 of the Committee of Ministers to Member States regulating the use of personal data in the police sector. Adopted by the Committee of Ministers on 17 September 1987.

Recommendation addresses issues such as the collection of data for police purposes, methods of storage, access restrictions and rights and independent overview as well as data security. The main principles are necessity, proportionality, lawful processing and purpose limitation.

At EU level, data protection is afforded by the Data Protection Directive[14]; however, Article 3(2) of that Directive takes data processing by public authorities outside its remit.[15] The Council Framework Decision 2008/977/JHA of 27 November 2008 (Framework Decision) addresses the processing of personal data for the purposes of police and judicial cooperation. It entered into force on 1 January 2009. The Framework Decision provides minimum standards to be maintained when processing personal data for the purposes of preventing, investigating, detecting or prosecuting criminal offences or executing criminal penalties of data which have been transmitted or made available between member states. Therefore, if the data were being shared with a police force in a different member state then this would apply, however the scope does not cover domestic processing of personal data by the competent judicial or police authorities in member states. Framework Decisions are similar to Directives in that they are binding as to the results to be achieved, but do not have direct effect.[16] This leaves the situation that there is currently no binding European legislative instrument specifically addressing the processing of data for policing purposes. However, even though the Data Protection Directive does not directly apply, the principles contained therein, which derive from Convention 108, still need to be applied to policing purposes.

As outlined in the legal framework in a previous chapter, data protection laws at European level are being updated and policing purposes are now covered by legislation. The new EU laws are the General Data Protection Regulation[17] (GDPR) and the Policing Directive,[18] which are aimed at strengthening citizens' rights while

---

[14] Directive 95/46/EC of 24 October 1995 on the protection of individuals with regard to the processing of personal data and on the free movement of such data [1995] OJ L 281/31 (Amended by Regulation (EC) of 29 September 2003 1882/2003 adapting to Council Decision 1999/468/EC the provisions relating to committees which assist the Commission in the exercise of its implementing powers laid down in instruments subject to the procedure referred to in Article 251 of the EC Treaty [2003] OJ L 284/1.

[15] Article 3(2)—'This Directive shall not apply to the processing of personal data: …in any case to processing operations concerning public security, defence, State security….and the activities of the State in areas of criminal law'.

[16] The UK did not legislate to bring the Policing Framework Decision into effect, but issued administrative circulars.

[17] European Parliament legislative resolution on the proposal for a regulation of the European Parliament and of the Council on the protection of individuals with regard to the processing of personal data and on the free movement of such data (General Data Protection Regulation) (COM(2012)11 – C7-0025/2012 – 2012/0011(COD)) Brussels 25.1.2012. Available at: http://eur-lex.europa.eu/legal-content/EN/TXT/PDF/?uri=CELEX:52012PC0011&from=EN.

[18] Proposal for a Directive of the European Parliament and of the Council on the protection of individuals with regard to the processing of personal data by competent authorities for the purposes of prevention, investigation, detection or prosecution of criminal offences or the execution of criminal penalties, and the free movement of such data (COM(2012)10 final, Brussels 25.1.2012.

reducing burdens for public authorities. These were both finally approved in April 2016, and Member States will have until 2018 to comply. The modernisation of the CoE's Convention 108 will continue in 2016 and will compliment and reinforce the new EU laws. The diagram in Fig. 11.2 illustrates the evolution of data protection laws at European level.

11.5   Applying the Law

Any processing of personal data constitutes an interference with human rights and would need to be justified. In doing so, any law enforcement authority would have to consider the relevant requirements of the legal instruments set out earlier, as well as the overriding data protection principles that, although not directly applicable in this context, must be adhered to. This approach has been endorsed recently in respect of the new GDPR and Police Directive.[19]

The key data protection principles[20] can be summarised as follows:

1. Fair and lawful processing
2. Purpose limitation
3. Data minimisation
4. Data retention

To ensure fair and lawful processing, any police action must be in accordance with the law and part of a legitimate aim pursued, but also necessary in a democratic society.[21] Purpose limitation and data minimisation are distinct but related principles: the first controls the reason for which certain data is being processed and the second specifies that the minimum amount of data to achieve the purpose can be processed.[22] The European Court of Human Rights (hereafter ECtHR) has consistently held that the storing and retention of personal data by public authorities interferes with Art 8 ECHR. Such interferences must have a legal foundation and be justified,[23] must include sufficient and relevant reasons, have a clear link with the

---

Available at: http://eur-lex.europa.eu/legal-content/EN/TXT/PDF/?uri=CELEX:52012PC0010&from=EN.

[19] The Article 29 Data Protection Working Party. (2015). Opinion 03/2015 on the draft directive on the protection of individuals with regard to the processing of personal data by competent authorities for the purposes of prevention, investigation, detection or prosecution of criminal offences, or the execution of criminal penalties and the free movement of such data, 3211/15/EN, WP 233. Adopted on December 1, 2015.

[20] See Article 6 of the Data Protection Directive and Article 5 of Convention 108 for data protection principles.

[21] Article 8(2) ECHR.

[22] See Article 6(b) and 6(c) of the Data Protection Directive, respectively.

[23] ECtHR, Leander v. Sweden, No. 9248/81, 26 March 1987.ECtHR, M.M. v. the United Kingdom, No. 24029/07, 13 November 2012.

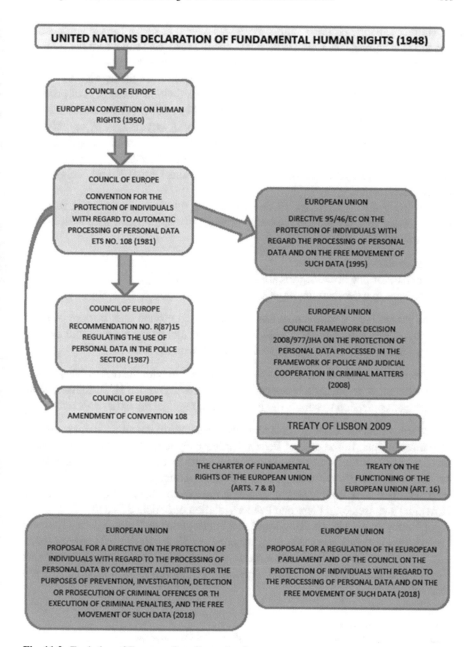

Fig. 11.2 Evolution of European Data Protection Laws

pressing social need and be proportionate. The retention of personal data by the police, particularly of non-suspects, is an area of concern.

Personal data that has been collected for one purpose (sharing information during a crisis) cannot be lawfully retained and shared with another data controller for an entirely separate purpose, unless there was a legal basis for the new processing and the data subjects were informed of the second processing in a sufficiently clear and precise way that an informed decision could be made by them. A legal basis other than consent might consist of one of the derogations in the legislation.

Article 8 of the Charter provides protection for individuals in respect of processing their personal data. The derogation provided in Art 52 (1) would be the one which would allow this right to be limited but only if provided for by law and the essence of the stated rights and freedoms is preserved. The principles of proportionality and necessity are emphasized and were reiterated recently by the Court of Justice of the European Union in the Schrems[24] and Digital Rights Ireland and Others[25] judgments.

In respect of the ECHR, the ECtHR has set out three criteria that must be satisfied to ensure that any interference with privacy is in compliance with Article 8(2) and must be:

1. in accordance with the law;
2. in pursuit of one of the legitimate aims in 8(2); and
3. necessary in a democratic society.

The jurisprudence in this area has all been based around one or more of these tests and is well established. In MM v United Kingdom[26] the court set out the following criteria for the action to be in accordance with the law:

1. have some basis in domestic law and be compatible with the rule of law;
2. the law must be adequately accessible and foreseeable that is, formulated with sufficient precision to enable the individual to regulate his or her conduct.

To be in pursuant of a legitimate aim means that one of the aims set out in Article 8(2) must be met. In the case of Peck v United Kingdom[27] the police were in possession of CCTV footage that had been used for an investigation into an attempted suicide. This was passed to the media who then publicised the footage. The court held that there were no relevant or sufficient reasons for this disclosure without the individual's consent and there had been a violation of Article 8 ECHR.

In Handyside v United Kingdom[28] the court said that '.....*necessary was not synonymous with indispensable...neither has it the flexibility of such expressions as admissible, ordinary, helpful, reasonable or desirable...*'[29] In the Sunday Times case[30] the court said that necessity shouldn't be interpreted too broadly nor too narrowly.

---

[24] Case C-362/14, Maximillion Schrems v Data Protection Commissioner.

[25] Joined cases C-293/12 and C-594/12, Digital Rights Ireland and Seitlinger and Others.

[26] MM v United Kingdom Appl. No. 24029/07 (ECtHR 13 November 2012).

[27] ECtHR, Peck v. the United Kingdom, No. 44647/98, 28 January 2003.

[28] Handyside v United Kingdom Appl. No. 5493/72 (ECtHR 7 December 1976).

[29] Ibid., para 48.

[30] The Sunday Times v United Kingdom Appl. No. 6538/74 (ECtHR 6 November 1980).

Any measure that interferes with an ECHR right should go no further than needed to fulfil the legitimate aim being pursued. In two cases[31] considering proportionality, the court accepted legitimate aim of the prevention or detection of crime or disorder but then asked whether it was necessary in a democratic society.

The Court of Justice (CJ) will also apply necessity and proportionality test to Articles 7 & 8 of The Charter, which will be read together. In the Schwarz case,[32] it was reiterated that limitations to fundamental rights must:

1. be provided for by law;
2. respect the essence of those rights;
3. be in accordance with the principle of proportionality, be necessary; and
4. genuinely meet objectives of general interest recognised by the Union or the need to protect the rights and freedoms of others.

The court said it must establish whether the limitations placed on those rights are proportionate to the aims and to the objectives. It will look at whether the measures implemented are appropriate for attaining those aims and not go beyond what is necessary to achieve them. The Article 29 Working Party[33] (hereafter WP29) have emphasised the importance of the concepts of necessity and proportionality when interfering with human rights in relation to processing personal data.[34] The WP29 provides practical guidance to LEAs and state that thought should be given to:

- the legal basis for the measure, particularly under Article 8(2) ECHR;
- the specific issue to be tackled such as the seriousness of the issue and social and cultural issues;
- the reasons behind the measure which are closely linked to decisions about data retention, minimised collection and data quality; and
- providing sufficient evidence to support the reasons for choosing the measure.

The ECtHR has set itself three tests when determining whether a measure is 'necessary in a democratic society'. The following criteria would be useful when considering whether an action is lawful:

1. pressing social need
2. proportionality—interference proportionate to legitimate aim
3. relevant and sufficient reasons.

---

[31] S & Marper v United Kingdom, Appl. No. 30562/04 (ECtHR 4 December 2008); Z v Finland, Appl. No. 2209/93 (ECtHR 25 February 1997).

[32] Schwarz v Stadt Bochum, ECJ, C-291/12 (CJEU 17 October 2013).

[33] The Article 29 Data Protection Working Party was set up under the Directive 95/46/EC of the European Parliament and of the Council of 24 October 1995 on the protection of individuals with regard to the processing of personal data and on the free movement of such data. It has advisory status and acts independently. http://ec.europa.eu/justice/data-protection/article-29/index_en.htm.

[34] Article 29 Data Protection Working Party (2014) Opinion 01/2014 on the application of necessity and proportionality concepts and data protection within the law enforcement sector, 536/14/EN, adopted 27 February 2014. Accessed 3 January 2016. Available at: http://ec.europa.eu/justice/data-protection/article-29/documentation/opinion-recommendation/files/2014/wp211_en.pdf.

In the UK, all law enforcement bodies are subject to the Data Protection Act. As such the Information Commissioner's Office[35] took formal action against a police force that used ANPR (Automatic Number Plate Recognition) for all vehicles in and out of a small town. Although the data retention standards were being met, the purpose was a policing one and no irrelevant or inaccurate data were processed, the ECtHR held the processing was unlawful because the force had failed to show a sufficiently pressing social need to justify the intrusion into so many innocent lives. The collection of data by technical surveillance or other automated means should also be based on specific legal provisions.

In terms of 'pressing social need' the jurisprudence points to the following considerations:

- Is the measure seeking to address an issue which, if left unaddressed, may result in harm to or have some detrimental effect on society or section of society?
- Is there any evidence that such a measure may mitigate such harm?
- What are the broader views of society on the issue in question?
- Have any specific views/opposition to a measure or issue expressed by society been sufficiently taken into account?

Each case will depend on its own merits. Whether the processing will satisfy the legal requirements will depend on many things but following the principles outlined earlier, as a minimum, should be the priority before any processing takes place.

Each crisis, each crime and each investigation or prosecution will be unique, so prescriptive advice is not realistic or appropriate. However, in terms of legal instruments, it can be seen that a clear picture is being created for the future; one with a more even landscape and with greater certainty on the horizon. The principles and the core values are being emphasised by both the CoE and the EU and are being upheld by the CJEU and ECtHR. By following these standards as a minimum, European LEAs might better achieve the thus far elusive balance, which is the theme of this chapter.

11.6   Issues Relating to ATHENA

In pursuing its objectives, ATHENA will necessarily create a complex series of legal relationships: relationships between contributors inter se, between contributors and the State, and also between contributors and their communications service providers, their employers, third sector crisis responders, potential litigants in criminal or civil proceedings, news media broadcasters, etc. These relationships will be potentially problematic unless appropriately identified and catered for right from the start. Further, if there is to be additional realisation of intellectual property rights within the products and outputs, it will be critical for ATHENA to have addressed all relevant legal issues arising from the creation of these complex relationships.

---

[35] Independent data protection regulatory authority in the UK. See: https://ico.org.uk/.

When it comes to the retention and processing of personal data, the LEA 'rap sheet' is arguably as relevant as anything the law might have to say on the subject. The level of public mistrust—particularly in the aftermath of the Snowden revelations[36]—ought to be as important to ATHENA as the substantive legal issues. This is because ATHENA's approach is entirely dependent on the establishment and development of trusting relationships between the relevant crisis responders.

The dilemma faced by LEAs when it comes to balancing criminal investigation/ prosecution and protection of the rights of citizens is nothing new,[37] but this dilemma has become more challenging with the arrival and development of Big Data capabilities and is possibly at its most acute in the field of the retention of personal data. In one of the examples referred to,[38] the police retention of DNA samples of individuals arrested, but later acquitted or had the charges against them dropped, was held to be a violation of the data subject's right to privacy.

Conversely, the failings of the police in England and Wales to retain relevant personal data in a searchable shared way so as to enable the tracking of dangerous offenders such as Ian Huntley[39] were widely reported and criticised in the Bichard Report[40] leading to wholesale changes in the UK police approach to operational IT capabilities. Data processing can all too easily be casually cast as mere 'bureaucratic' compliance,[41] and public and political tolerance of administrative niceties when faced with preventable criminality can be expected to be unforgiving of the LEAs involved. However, the link between police legitimacy and trust of their communities—particularly when it comes to use of intrusive powers and data processing—is too significant for ATHENA to ignore.

Taken together with a degree of growing global mistrust of State use (and abuse) of personal data[42] and the development of what some have seen as a pervasive 'omniveillance' made possible by Big Data [22], the need to ensure transparency and legitimacy at all stages ought to be a cornerstone of its settings.

In addition, other criminal justice services—such as those offered to victims of crime extended in compliance with the Victims' Code[43] in accordance with published guidance from the UK's Office of the Information Commissioner[44]—are

---

[36] See: http://www.theguardian.com/world/the-nsa-files.

[37] This public interest dichotomy was captured by Lord Bingham at 65–66 of *R v Lewes Crown Court ex parte Hill* (1991) 93 Cr App R 60.

[38] ECtHR, S. and Marper v. the United Kingdom, Nos. 30562/04 and 30566/04, 4 December 2008.

[39] Convicted on 17 December 2003 of the murder of 10-year-old schoolgirls Holly Wells and Jessica Chapman.

[40] Report of the Bichard Inquiry HC 653 22 June 2004, The Stationery Office, London.

[41] See e.g.: http://www.dailymail.co.uk/news/article-449456/Paper-tigers-lunatic-bureaucracy-crippling-police.html.

[42] See: http://www.theguardian.com/world/the-nsa-files.

[43] See the specific website set up by the Police and Crime Commissioner for West Yorkshire: www.helpforvictims.co.uk.

[44] See: https://ico.org.uk/media/for-organisations/documents/1068/data_sharing_code_of_practice.pdf.

beginning to focus on the data control and sharing arrangements. It is therefore probably time that data control protocols were built in to all State agencies' policies as a standard and ATHENA might make a useful contribution to this as a by-product.

When looking at the practicalities of how ATHENA should address these issues, it is worth looking beyond the criminal justice setting. Although having a particular LEA frame of reference, in some respects the legal data considerations created by ATHENA reactions are similar to those affecting commercial relationships [23], making the 'customer' a fully empowered actor in the market place, rather than one whose power is entirely dependent on exclusive relationships—in this case with the State and its agencies rather than commercial vendors—particularly if those relationships are based on coerced agreement. However, a photo-sharing policy for a non-investigative agency appears relatively simple and very different from sharing with LEAs that have investigatory duties, powers and processes.[45]

One suggestion for how to manage the specific LEA-based stoichiometry of ATHENA would be to borrow from the commercial sector and create an End User Agreement Licence (EUAL) between ATHENA participants and the LEAs/State agencies in receipt of the data. Following the same principles as those being promoted in the context of commercial data exchange, such an EUAL would make ATHENA transactions 'bidirectional' [24].

Should an LEA acquire personal data in the course of an ATHENA-related crisis (say, a civil contingency such as a flood) that will be potentially relevant to the investigation of subsequent criminal investigation (offences of looting) the temptation (or arguably obligation) for those agencies to retain those data beyond the time of the exigencies of the rescue/responder requirement, will often be irresistible. ATHENA-based data can—and in fact are designed to—produce specificity in key elements such as the time, identity and location of the contributor. While the value of such data in the course of the combined effort to neutralize the threat, risk and harm of the presenting crisis is self-evident, so too is the correlative value of those data to other—perhaps unrelated—investigations or simply intelligence. How far the participants can be taken to have consented to the retention and use of their personal data for divergent purposes will be important within the legal framework and ought, therefore, to be addressed at the point of recruiting ATHENA citizens.

11.7 Conclusion

Public trust is arguably a sine qua non of any public engagement in the way envisaged—and indeed relied upon—by ATHENA. Any generally applicable issues of public trust around crises[46] are clearly made more acute by the involvement of LEAs

---

[45] For example see: International Committee of the Red Cross (ICRC) at: http://www.flickr.com/photos/ifrc/sets/72157623207618658/; See also: Maple Bluff [Wisconsin] Fire Department at: http://picasaweb.google.com/MapleBluffFireDepartment; See also Virginia Department of Emergency Management: http://www.flickr.com/photos/vaemergency/.

[46] For a discussion of the public's pre- and post-disaster trust of social media, engagement during disasters and behaviour and attitude change, see: Jin & National Liu (2010); Murdogh (2009); and Hagar (2013).

who have coercive and intrusive powers. As such, an obvious caveat to ATHENA in this regard is that the remote utilisation of private social relationships forged by the reactions to crisis comes, if not to be used, then at least to be suspected by communities as another form of surveillance.

Given that the Council of Europe has expressed deep concerns on the legal implications of mass surveillance revealed by Edward Snowden and the correlative unlawful State use of personal data accumulated by private businesses,[47] and given too that the Council has concluded that mass surveillance by LEAs has been ineffective in preventing terrorism,[48] it would be wise for ATHENA expressly to disavow any general surveillance purpose at the outset and to provide undertakings in relation to the further processing of personal data. Given also the concerns over State surveillance of public areas more generally[49] and the ongoing controversy around statutory powers for State interception of data and intrusive tactics,[50] it would be wise to address the very real risk that LEA usage of ATHENA Big Data might be seen as an extension of State surveillance—and a covert, unregulated one at that.

ATHENA aims ambitiously and pragmatically to harness the 'collective problem solving' [25] of citizens using social media while, at the same time, developing 'Europe-wide and internationally transferable guidelines for protocols, systems, technologies, techniques and good practice in the use of new communication media by the public to increase the security of citizens in crisis situations'.[51] These 'guidelines' must necessarily include a clear data protocol (possibly in the form of an End User Agreement License) to protect the security of citizens' personal data, identities and privacy and to safeguard the relationships that are critical to ATHENA's scalability.

The EU and the CoE have consistently stressed the importance of fundamental rights in relation to the digital age and any interference with them must be justified and lawful, satisfy the principles of necessity and proportionality in particular, and also adopt a citizen's perspective. After all, joining in the conversation on social media platforms is not a right that the police have; it is only possible if the general public consent to it, it is based on the will and acceptance of the citizens and if this good will and acceptance goes, the police will have nothing. The police are facing an uphill struggle in the face of increasing concerns about surveillance and data processing by public authorities. The public are more aware than ever of their rights and more determined than ever to give effect to them. Consent and good will are key elements of any successes in this area.

---

[47] Council of Europe Committee on Legal Affairs and Human Rights Draft Resolution and Recommendation adopted 26 Jan 2015.

[48] Council of Europe Resolution 2031 (2015) Terrorist attacks in Paris: together for a democratic response para 14.2.

[49] See Council of Europe Doc. 11692 21 July 2008 Video surveillance of public areas Recommendation 1830 (2008) Reply from the Committee of Ministers adopted at the 1032nd meeting of Ministers' Deputies (9 July 2008).

[50] See e.g.: Liberty report on second reading of Counter-terrorism and Security Bill House of Commons December 2014.

[51] ATHENA submission, document B.1.1.12 para 4.

It can be seen that the main difficulty is finding the appropriate balance between respect for personal privacy and securing the safety and the welfare of the community at large. This dilemma has been illustrated in a number of ways and although several battles have been fought, the war continues. Weighing up public interest against State interests is an almost impossible task and will inevitably be decided on a case-by-case basis. What is clear though is that there are key principles that run through all the legislative instruments in this area: transparency, proportionality, necessity, data quality, data minimisation and purpose limitation. These have been repeated and will certainly feature in the forthcoming EU Directive. Both the ATHENA system and the use of social media by the police have one thing in common; they rely upon cooperation by the public.

There are many positive aspects to police use of social media, but just as this aspect of public life has changed, so must the way the police operate. This is true policing by consent and great care must be taken to achieve the correct tone and balance. Just as social media is a powerful resource for law enforcement agencies, so it is a powerful tool for uniting people the world over, and any disrespectful or overly intrusive methods of such policing has the potential to shut many virtual doors. It is clear that both the EU and the Council of Europe have digital security at the heart of their core principles; this issue isn't going to go away and will not remain unnoticed. The police need to get it right and ensure transparency, legitimacy and integrity. The public and the law makers will stand for nothing less.

References

1. Patrick Meier Digital Humanitarians. (2015). CRC Press, Taylor & Francis Group; Wendling, C., Radisch, J., & Jacobzone, S. (2013). The use of social media in risk and crisis communication. *OECD Working Papers on Public Governance*. No. 24. OECD Publishing. Accessed 9 September 2015. Retrieved from http://dx.doi.org/10.1787/5k3v01fskp9s-en; Waton, H., et al. (2014). Citizen (in)security?: Social media, citizen journalism and crisis response. In *Proceedings of the 11th International ISCRAM Conference*, University Park, PA.
2. Akhgar, B., & S. Yates (Eds.). (2013). *Strategic intelligence management*. Butterworth-Heinemann. Retrieved from http://dx.doi.org/10.1016/B978-0-12-407191-9.00015-6.
3. UK Information Commissioner's Office (ICO). (September 2014). *Annual track 2014*. Retrieved March 30, 2016, from https://ico.org.uk/media/about-the-ico/documents/1043485/annual-track-september-2014-individuals.pdf; Special Eurobarometer 431, Data protection. Retrieved from http://ec.europa.eu/public_opinion/archives/eb_special_439_420_en.htm#431.
4. Crump, J. (2011). What are the Police Doing on Twitter? Social Media, the Police and the Public. *Policy and Internet, 3*(4), article 7.
5. Smadja, F. (1993). Retrieving collocations from text: extract. *Computational Linguistics, 19*(1):143—177; Lin, D. (1998). Extracting collocations from text corpora. In *First Workshop on Computational Terminology* (pp. 57–63), Montreal, Canada; Seretan V., Nerima, L., & Wehrli, E. (2003). Extraction of multi-word collocations using syntactic bigram composition. In *Proceedings of International Conference on Recent Advances in NLP*. Issue: Harris 51. Publisher: Citeseer.
6. Sampson, F. (2015). Cybercrime presentation Project COuRAGE and CAMINO Cyber Security Workshop, Montellier, France.

7. The Police Foundation. (2014). *The briefing: Police use of social media.* Retrieved from http://www.police-foundation.org.uk/publications/briefings/police-use-of-social-media.

8. Bayerl, P. S. (2012). Social media study in European Police Forces: First results on usage and acceptance. *COMPOSITE project.* Retrieved from www.composite-project.eu.

9. Accenture. (2014). *How can digital police solutions better serve citizens' expectations?* Retrieved March 30, 2016, from https://www.accenture.com/gb-en/~/_acnmedia/Accenture/Conversion-Assets/DotCom/Documents/Local/en-gb/PDF_2/Accenture-How-Can-Digital-Police-Solutions-Better-Serve-Citizens-Expectations.pdf#zoom=50.

10. Kotronaki, L., & Seferiades, S. (2012). Along the pathways of rage: The space-time of an uprising. In S. Seferiades, & H. Johnston (Eds.), *Violent protest, contentious politics, and the Neoliberal State* (pp. 159–170). Surrey: Ashgate; Russell, A. (2007). Digital communication networks and the journalistic field: The 2005 French riots. *Critical Studies in Media Communication, 24*(4): 285–302; Kavanaugh, A., Yang, S., Li, L., Sheetz, S., & Fox, E. (2011). Microblogging in crisis situations: Mass protests in Iran, Tunisia, Egypt. *CHI2011*, Vancouver, Canada, May 7–12, 2011; Papic, M., & Noonan, S. (February 3, 2011) *Social media as a tool for protest. Security Weekly.* Retrieved December 10, 2014, from http://www.stratfor.com/weekly/20110202-social-media-tool-protest#axzz3LWjMNk4d; Xiguang, L., & Jing, W. (2010). Web-based public diplomacy: The role of social media in the Iranian and Xinjiang Riots. *The Journal of International Communication, 16*(1): 7–22.

11. NPIA. (2010). *Engage: Digital and social media for the police service.* London: National Policing Improvement Agency.

12. Casilli, A., & Tubaro, P. (2012). Social media censorship in times of political unrest—A social simulation experiment with the UK riots. *Bulletin de Methodologie Sociologique, 115*, 5–20; Howard, P., Agarwal, S., & Hussain, M. (August 9, 2011). *When do states disconnect their digital networks? Regime responses to the political uses of social media.* Retrieved November 25, 2014, from http://ssrn.com/abstract=1907191.

13. McSeveny, K., & Waddington, D. (2011). Up close and personal: The interplay between information technology and human agency in the policing of the 2011 Sheffield Anti-Lib Dem Protest. In B. Akhgar & S. Yates (Eds.), *Intelligence management: Knowledge driven frameworks for combating terrorism and organized crime* (pp. 199–212). New York: Springer.

14. Liberty. (2011). *Liberty's report on legal observing at the TUC March for the alternative.* Retrieved November 22, 2014, from https://www.liberty-human-rights.org.uk/sites/default/files/libertys-report-on-legal-observing-at-the-tuc-march-for-the-alternative.pdf.

15. NETPOL-Network for Police Monitoring. (2011). *Report on the policing of protest in London on 26 March 2011.* Retrieved November 22, 2014, from https://netpol.org/wp-content/uploads/2012/07/3rd-edit-m26-report.pdf.

16. Denef, S., Kaptein, N., Bayerl, P., & Ramirez, L. (2012). Best practice in police social media adaptation. *COMPOSITE project,* Procter, R., Crump, J., Karstedt, S., Voss, A., & Cantijoch, M. (2013). Reading the riots: What were the police doing on Twitter? *Policing and Society: An International Journal of Research and Policy, 23*(4), 413–436.

17. Denef, S., Kaptein, N., Bayerl, P., & Ramirez, L. (2012) Best practice in police social media adaptation. *COMPOSITE project.*

18. Procter, R., Crump, J., Karstedt, S., Voss, A., & Cantijoch, M. (2013). Reading the riots: What were the police doing on Twitter? *Policing and Society: An International Journal of Research and Policy., 23*(4), 413–436.

19. Beguerisse-Díaz, M., Garduno-Hernandez, G., Vangelov, B., Yaliraki, S., & Barahona, M. (2014). *Interest communities and flow roles in directed networks: The Twitter network of the UK riots.* Cornell University Library. Retrieved from http://arxiv.org/abs/1311.6785; Bruns, A., & Burgess, J. (2012). *#qldfloods and @QPSMedia: Crisis Communication on Twitter in the 2011 South-East Queensland Floods.* ARC Centre of Excellence for Creative Industries and Innovation: Queensland University of Technology, Brisbane, QLD.

20. Loveys, K. (November 11, 2010). Come down from the roof please, officers tweeted. *Mail Online.* Retrieved from http://www.dailymail.co.uk/news/article-1328586/TUITION-FEES-PROTEST-Met-chief-embarrassed-woeful-riot-preparation.html.

21. Procter, R., Crump, J., Karstedt, S., Voss, A., & Cantijoch, M. (2013). Reading the riots: What were the police doing on Twitter? *Policing and Society: An International Journal of Research and Policy, 23*(4), 413–436.
22. Blackman, J. (2008). *Omniveillance, Google, privacy in public, and the right to your digital identity: A tort for recording and disseminating an individual's image over the Internet.* 49 Santa Clara L. Rev. 313; Armstrong, T, Zuckerberg, M, Page, L., Rottenberg, E., Smith, B., & Costelo, D. (December 9, 2013) *An Open Letter to Washington.*
23. Searls, D. (2012). *The intention economy: When customers take charge.* Cambridge, MA: Harvard University Press.
24. Lanier, J. (2013). *Who owns the future?* New York, NY: Simon and Schuster.
25. Palen, L. (2008). On line social media in crisis events. *Educause, 3,* 76–78; See also: G. Baron. *Social Media and Crisis: A Whole New Game.* Retrieved from http://www.youtube.com/watch?v=MFt7NXDhcmE; Vieweg, S., Palen, L., Liu, S., Hughes, A., & Sutton, J. (2008). Collective intelligence in disaster: An examination of the phenomenon in the aftermath of the 2007 Virginia Tech Shootings. In *Proceedings of the Information Systems for Crisis Response and Management Conference (ISCRAM 2008).*

Part IV
Testing and Evaluating the ATHENA System

Chapter 12
Preliminary ATHENA Case Studies: Test-Bed Development and Delivery

Julij Jeraj, Andrej Fink, Alison Lyle, Tony Day, and Kevin Blair

12.1 Introducing the Case Studies: Their Purposes and Objectives

In order to ensure the successful delivery of the Athena system, an essential stage of the development process involves the testing and validation of its components, functionality and usability at various key stages to ensure that its technical capabilities are on track and that they match up to the expected user needs and requirements. The exercises described in this chapter were designed and implemented in order to carry out the testing in a robust and thorough manner. The design, presentation, format and scenario of each exercise were varied with the intention of producing progressive technological improvements by incorporating a number of potential scenarios, providing an enhanced testing regime as the complexity increased. The lessons accruing from these cases were then carried forward in a final test application of the ATHENA system, to be reported in the next chapter of this volume.

J. Jeraj (✉)
The City of Ljubljana, Emergency Management Department, Ljubljana, Slovenia
e-mail: julij.jeraj@ljubljana.si

A. Fink
University Medical Centre Ljubljana, Ambulance Service, Ljubljana, Slovenia
e-mail: andrej.fink@kclj.si

A. Lyle
Police National Legal Database, Wakefield, UK
e-mail: Alison.Lyle@westyorkshire.pnn.police.uk

T. Day
CENTRIC, Sheffield Hallam University, Sheffield, UK
e-mail: T.Day@shu.ac.uk

K. Blair
Office of the Police and Crime Commissioner for West Yorkshire, Wakefield, UK
e-mail: Kevin.Blair@westyorkshire.pnn.police.uk

© Springer International Publishing AG 2017
B. Akhgar et al. (eds.), *Application of Social Media in Crisis Management*,
Transactions on Computational Science and Computational Intelligence,
DOI 10.1007/978-3-319-52419-1_12

An important element of successful practical exercises is to engage the participants sufficiently in order to persuade them that they could be in the midst of a real crisis. Given that the way that people communicate in crisis situations and the means by which they understand information is very different to when they are in normal situations, it was particularly important for the ATHENA system, which is to be used in crisis situations, to be tested in such a way that tried to replicate the state of shock and panic they may experience within a crisis. Although situations of this nature are virtually impossible to replicate entirely, every effort was made to take this important factor into account when designing and writing the exercises involved.

Each of the three exercises had a different key emphasis. The first of the three exercises focused on the technological capabilities of the early prototypes of both the ATHENA mobile application (app) and ATHENA Crisis Command and Control Intelligence dashboard (CCCID). The main features of these two applications were put through a robust testing procedure, based around an earthquake scenario, which was carried out by consortium members in Izmir, Turkey, in May 2015.

Inherent in the second exercise (held in Ljubljana, Slovenia at the end of January 2016) was a greater focus on expanding and validating user requirements. This case involved a much larger group of participants, none of whom were consortium members. Rather, they were representatives of the various end-user groups who would be destined to use the system in real-life situations. Although technical elements were tested here, a far greater emphasis was placed on the system's fitness for purpose from an end-user perspective. The scenarios for these initial exercises were carefully selected so that several important factors could be incorporated and the testing could be regarded as having been as 'authentic' as humanly possible (see Chap. 5).

A large field exercise, also held in Ljubljana (at the beginning of April 2016), served as a test-bed for the third practical testing of the ATHENA products. This took place in a highly complex environment in which several terrorist actions resulted in a mass casualty accident, hostage situation, fire and a structural collapse in a fully packed football stadium, hosting with an international match that was also watched by VIPs from the two rival nation states. In this case, ATHENA was participating as part of a larger multi-agency exercise.

The nature, rationale and other relevant details of the three separate exercises are now set out in turn. Due to its considerably more complex character, the third exercise is dealt with much more expansively than its two predecessors. Nevertheless, the lessons arising from the three cases and the ways in which these have been fed into the ongoing development of the ATHENA system are dealt with in equal measure in the concluding section of the chapter.

12.2 The First Exercise: Izmir (Turkey)

The first exercise was hosted by the Fire Service of Izmir and took place at their Crisis Management Centre just outside the city. An earthquake scenario formed the basis of this first exercise. This built on the theme of earlier work in the project and provided a culturally appropriate backdrop for the storylines and scripts. The whole exercise was heavily scripted to avoid decision-making by the participants: a book was provided to each that contained their 'story' and a series of specific reports to be sent at certain times from pre-specified locations. Three different locations were used which allowed the writers to include different geographical features and environmental hazards. These incorporated a range of incidents and challenges on which to base the scripts. After the orientation phase, participants were transported to three different locations around Izmir to carry out the main part of the exercise which took approximately 3 h from start to finish.

The design of the first exercise was based entirely on the existing functional user requirements of the system for that stage of the project. These were established during a workshop session attended by consortium members and external participants (with an experience of crisis management) at an earlier stage of the project's evolution. The exercise was divided into four distinct phases to allow for the testing of various elements, as well as to simplify and clarify the process, both from the perspectives of those taking part and those charged with evaluating and facilitating the event. Putting it more specifically, the phases were as follows:

1. Initial use and familiarisation with the App and scenario
2. Deployment of participants to various locations around Izmir
3. Sending and receiving reports
4. Evaluation and debrief

The sending of messages was split into two phases. The first phase was designed to represent the period of time immediately following an earthquake when a large number of reports were likely to reflect the fear and panic generated by such a disaster. The second section was designed to approximate to a recovery phase, with fewer reports being sent and a greater number of tasks emerging for participants to carry out using data from the first phase.

Nine participants from the ATHENA consortium acted as app users (see Chap. 7) in the exercise, while another member acted as the CCCID operator. Each App user was assigned a 'role' and provided with personal details and a background story which included family members, places of work and life circumstances. This was designed to encourage participants to become as thoroughly engaged in the exercise as possible and to put the use of the app into a seemingly realistic context. Furthermore, the packs also included details of a number of tasks to be carried out by each participant at particular points of the proceedings; for example, sending specific messages at a specific time, searching the app for information, or uploading videos and photographs. This tied in with the actions from other participants and the CCCID operator, thus helping to simulate a real crisis situation (see Chap. 5).

In total, 108 pre-prepared written reports were made available for the App participants to send in the course of the exercise; a further 25 reports were posted in advance of the exercise in order to increase the amount of information present in the system, and the CCCID operator posted an additional 12 reports while the exercise was in progress that were designed to respond to the messages sent in by the app users. Figures 12.1 and 12.2 show the app with a number of uploaded reports that

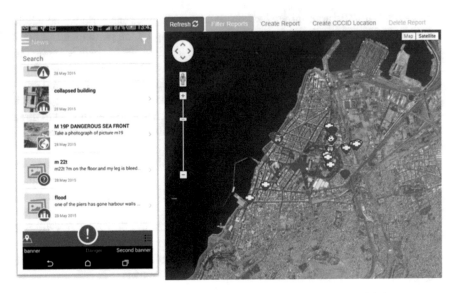

Fig. 12.1 Reports uploaded to the ATHENA app during the Izmir exercise and their appearance on the CCCID

Fig. 12.2 A user accessing reports via the crisis map on the phone and a CCCID user accessing a report through the dashboard including important location (*blue*) pins

were available to view for app users and the corresponding map view in the CCCID. The CCCID controller remained at the Crisis Management Centre throughout, and all those involved returned at the end of the day for a feedback and debriefing session. Facilitators and observers took notes throughout in order to collect as much feedback as possible. This feedback fed into the future development of the Athena system and is reported in Sect. 12.5.

12.3 The Second Exercise (Ljubljana, Slovenia)

The second exercise, held in Ljubljana, Slovenia, was hosted by the Athena partners the Municipality of Ljubljana and Fire Brigade Ljubljana. The scenario underpinning the exercise (which was conducted some 8 months after its predecessor) was partly predicated on an environmental disaster in which a large volume of poisonous sludge containing dangerous by-products from industrial processes leaked from a dam at high level and spread down the mountainside into urban areas. This particular aspect of the scenario provided an opportunity to involve a number of different types of end-users, namely: first responders, Non-Governmental Organisations (NGOs), citizens of various ages and social backgrounds and environmental agencies. It also allowed the writers to develop a progressive 'story' that facilitated the testing of appropriate technical elements while enabling users to become fully engaged and capable of providing feedback on their situational needs.

The second part of the scenario presupposed a developing situation in which the man-made environmental disaster referred to above resulted in a release of chemicals into a river with the consequential threat to potable water. This formed the basis for the strategic decision-making activities of appropriate organisations and their expert personnel, who had the capability to distribute and share relevant knowledge by using the ATHENA Decentralised Intelligence Processing Framework's Logic Cloud. This feature enables wider collaborative decision-making in crisis situations by distributing information and data collected by the ATHENA system to relevant experts and then feeding accurate information back to crisis managers in a timely fashion. This was the first test of this integrated element of the system (see Chap. 9).

The main operational hub of the second exercise was the Fire Service Headquarters in Ljubljana. All participants began the day at this location, receiving instructions and carrying out familiarisation tasks. Facilitators also received instructions and information to assist them in carrying out their roles. The app users then divided into pre-defined groups which moved to various locations to carry out the live exercise. This second exercise was much less prescriptive than the first and was designed in such a way as to allow greater autonomy and decision-making by participants, especially on the part of the app users. A conscious decision was made to use participants that were not directly associated with the project thus meaning their experience of the app and the project in general is not influenced by prior involvement. Books similar to the ones successfully used in the first exercise were provided, but contained fewer instructions relating to the content of reports, therefore requiring participants to make their own choices and decisions about what information and content to send via the app (see Chaps. 7 and 10). This allowed them to experience using the app and its features in a way that would replicate a real situation. All

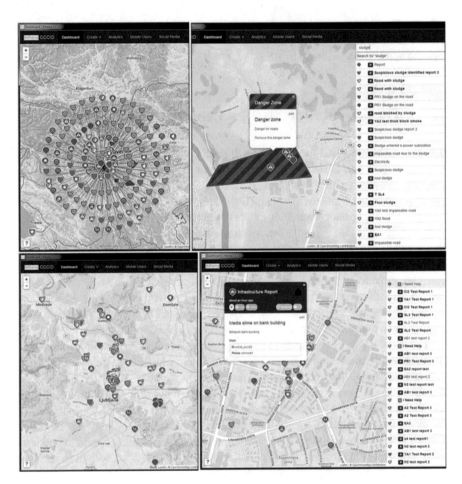

Fig. 12.3 The CCCID during the second exercise. Mapping danger zones, report pins and requests for help

reports were received into the CCCID, the distribution and examples of these reports are shown in Fig. 12.3.

Again, different geographical locations were employed; and because participants using the app were able to move between them in the course of the exercise, this helped facilitate the testing of geolocation features. Remote locations were also set up in the United Kingdom and Sweden where simultaneous table-top exercises were introduced. This was to allow the exercise to incorporate elements that would approximate the experience of the families and friends of the participants who were using information derived from the app and social media posts as vital sources of information about the unfolding crisis. For practical purposes, the facilitators of each were in contact with a coordinator present at the live exercise in Slovenia, via a video link. Role players for the social media feed were located in Sweden (Swedish National Defense College in Stockholm) and UK (Wakefield District Police Headquarters) who provided 5 role players each.

Corresponding role players for the app testing component of the exercise came from a range of organisations affiliated with Ljubljana's emergency response community and academia, most notably the city's Emergency Management Department, fire service (professional and volunteer staff), city police, scouting organisations, Cave Rescue Service, national police, ambulance service, and professors and students drawn from the Faculty of Criminal Justice and Security of the University of Maribor. Participants engaged in the testing of the Athena Logic Cloud were expert members of the city's emergency management department, the Department of the Environment, fire service (professional staff), members of a water company, the National Geological Survey Institute and private Water Science Institute. All told, there were 50 participants occupying various different roles in Ljubljana, and a further seven people acting as observers.

12.4 The Third Exercise (Ljubljana, Slovenia)

12.4.1 Background and Design

The content of the third ATHENA exercise was motivated by discussion that occurred in the aftermath of the terrorist attacks in Paris on 23 November 2015. In this case questions were raised among blue light services operating in the City of Ljubljana with regard to the level of their preparedness (and quality of coordination and co-operation between themselves and other relevant stakeholders) for a possible similar atrocity of this nature. The exercise was thus designed with a view to testing the efficiency and effectiveness of existing plans and procedures as part of a large emergency planning exercise within Slovenia. As part of this a commitment was also made to incorporate a related test trial of the ATHENA system, which sought to examine in particular:

1. The ways and degree to which individuals made use of the App in sending reports, and utilised information made available by the CCCID, and
2. The ease and efficiency involved in the way that individuals were able to receive, understand and process information from App users' reports and provide verified information to App users in return.

It was with these general objectives in mind that the major blue light services of Ljubljana accepted the invitation of the University Medical Centre, Ljubljana, to carry out a full-blown mass casualty field exercise, centering on the local national football stadium patient-receiving hospital. The exercise was based on a 'secret' scenario, i.e., one not to be disclosed beforehand to the emergency services, involving at least 200 role players (out of a total of several hundred more) who would incur serious injury.

The final scenario revolved around a football match between the Slovenian national team and that of a rival foreign country. The scenario unfolded as follows. A football match was attended by presidents and other high-profile dignitaries from both nations—a situation requiring high levels of security and protection—as well as a crowd of around 17,000 spectators, who were unprepared for the explosion that would occur in the south section of the stadium, 20 min into the game resulting in a large number of injuries and fatalities. Following the explosion, the Police VIP Protection Unit initiated the evacuation of the VIPs under their responsibility. However, while undertaking such action the officers involved found themselves being shot at by two assailants standing at the exit to the VIP box. This gunfire resulted in one VIP, several police officers and visitors being wounded, a minority of them fatally.

In the aftermath, the two assailants kidnapped several VIPs and confined them to one of the stadium's executive boxes where they then threatened to kill them. Shortly after this, the attackers remotely triggered a second bomb in another section of the stadium. This produced an instant fire, followed by a structural collapse which resulted in numerous further injuries. The spectators fleeing from the stadium were met by one of the armed assailants who started to open fire on them. This attacker was soon shot dead by police; but it took a further 3 h for the security forces to rescue the VIP hostages.

The multiple facets to this scenario were designed to pose significant challenges to the fire and medical services, whose efforts were hampered by these ongoing events. Mock injuries of differing severity were sustained by 240 participants, with another 489 role players also participating but not as casualties.

'Civilian' role players were mostly drawn from local academic institutions (e.g., nursing and criminal justice departments) and occasional volunteer organisations, such as a local boy-scout club. The security and emergency responses were carried out by 879 members of relevant blue light teams from Ljubljana and other districts of Slovenia. The ATHENA App was used by those playing role of football fans and some of the rescue personnel. Examples of the images submitted through the app can be seen in Fig. 12.4. The CCCID was set up in the central station (headquarters) of the Ljubljana Fire Service's main station (though the operation was led by a central leadership core, located in the stadium control room). The 'match' itself was contested by rival teams of local civil servants and Army personnel. 'Casualties' were initially taken to the stadium's medical centre; however, as the scenario played out, other victims were transported either to the trauma ward of the nearby University Medical Centre, or (depending on the severity of their injuries) to other hospitals in the surrounding area. Fire-fighting teams with a total of 33 emergency vehicles were drawn from 12 local stations.

12.4.2   Implementing the ATHENA System

Considerable energy was devoted to adapting the ATHENA CCCID and App to the requirements of this exercise. The app's interface was translated from English into Slovenian and Crisis Management Language (CML—method of automatically extracting information from report text into structured language) was also adapted in similar manner. In practical terms, the CCCID was operated via six computers set up in the central station of the Ljubljana Fire Service. CCCID operators and controllers

Fig. 12.4 Medical Service and medically trained volunteer members of Fire service jointly perform first triage and treatment while on other sections of the stadium injured role players still wait to for assistance (these photos are retrieved from the ATHENA system)

were drawn from Fire Service dispatch and commanding personnel, and the Police, and Emergency Management Department. Most of these individuals had been exposed to the tool 2 months earlier as part of the second ATHENA testing exercise. Facilitators provided additional training at a 4-h session on the day before the exercise.

A primary objective of this exercise was to gauge the extent to which individuals might be inclined to use the app in the absence of any substantial training or preliminary instructions, or of the type of scripted directions on what to do, how and when, especially since many of them had participated in the second exercise where they were given roles that were more formally scripted. Indeed, the only information issued to the majority (some 700) of the role-playing participants took the form of an e-mail drawing their attention to the link to the appropriate download page and very basic instructions on when, how and for what purpose they might use the app in the course of the exercise. This contrasted with the short presentation given 2 weeks prior to the exercise to some participants occupying specialist roles (such as police officers). Some of the role players, including about 100 from the Police Academy, were also informed about the ATHENA system and invited to download and test out the app.

It was according to these arrangements that we were able to test the effectiveness of the ATHENA system in relation to a large-scale exercise. In the course of this process, role players and responders using the app produced 122 reports and 55 help requests. These were received by five CCCID operators, assisted by specially designated facilitators and evaluators. Figure 12.5 show users submitting an image through the app's interface during the exercise while Fig. 12.6 shows the extent and distribution of the reports around the football stadium on the CCCID.

Fig. 12.5 Role players using ATHENA application to send photos (courtesy of Matic Macek)

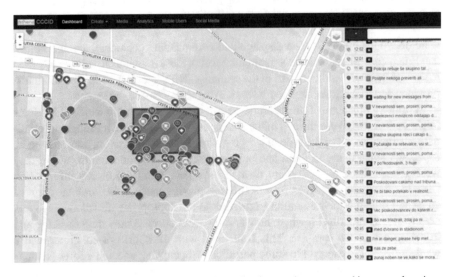

Fig. 12.6 Dashboard view of ATHENA mobile application user's reports, and important locations and danger zone added by Command and Control Intelligence Dashboard operators

12.5   Outcomes and Lessons Learned

All three of the preliminary exercises satisfied their main purpose: of contributing to the continued improvement and development of the ATHENA system in line with user requirements and expectations. Of particular importance to the ongoing development of ATHENA were the comments and suggestions from participants, which were duly incorporated in the evaluation reports and fed back to technical partners. None of the difficulties encountered by users were insurmountable. Indeed, most of these were dealt with on the day or addressed soon after the conclusion of the exercise and were often incorporated into the roadmap for the design and development of both the app and the CCCID.

The lessons learned from each of the first two exercises helped contribute to the iterative process of user-centred design, by which the ATHENA system has developed. For example, in the first exercise it was evident that a single CCCID operator would not be able to deal with all the incoming data in order for the system to operate at or near real-time. This was fed directly into the design of the second exercise which saw the CCCID role divided into five separate parts, each one concentrating on a separate function. This led, in turn, to the further important lesson—that it would be beneficial to have in place a large screen showing all operators the full picture of the crisis as it was unfolding. Another important lesson emerging from the first exercise was the need to have a 'setting-up day' prior to the exercise, in order to ensure efficient and effective functioning of all technical equipment, and a reduced possibility of something going wrong while the exercise was in full swing.

Some of the main issues for the CCCID in the first exercise related to refreshing the crisis map and validating incoming messages. However, these were overcome and the exercise progressed as planned. The app suffered some stability issues which were rectified by the later exercises. Both sets of problems provided valuable learning opportunities and neither of them prevented the exercise continuing as planned.

In the second exercise fewer major problems were encountered and all parts of the system worked well and were integrated. From a technical point of view, the system performed to its capabilities at that stage of development. Issues were minor and served to provide valuable information to technical development teams.

The second of the Ljubljana exercises (Exercise 3) was a valuable testing opportunity for the ATHENA system. Testing was carried out by the largest number of app users so far, the vast majority of whom were required to use it spontaneously reporting whatever events they could see happening at the time. CCCID operators and controllers were using the tool in an unscripted, independent way forcing them to deal with a large number of incoming reports without prior knowledge of how the incident would unfold. This meant they were able to fall back onto their professional judgment about how to use incoming data to assist emergency services and the affected population. Furthermore, newly introduced functionality could be tested. So, for example, this third exercise presented the consortium with the opportunity to test a specific piece of data processing functionality known as CML (crisis management language). CML enables the extraction of particular terms from a report's text that facilitates the understanding of that message's content allowing it to be automatically translated. Of the 112 reports received by CCCID, 42 were understood by the prototype CML component; that is particular terms could be extracted. The translation of report text into CML text was practically instantaneous; however, among the most valuable lessons to be learned was the need to take account of the tendency for users to send reports by entering the most important information in the subject field rather than the body of the report, and incorporating this in the scan that activates the translation module. Character coding issues also

J. Jeraj et al.

became apparent in the post-exercise analysis, providing an opportunity for this to be addressed in ongoing development.

All components of the ATHENA system operated as reasonably as could have been expected given their technological immaturity. Tens of users in the exercise play area engaged well with the app, whilst in the Ljubljana fire brigade headquarters, six operators and controllers validated incoming reports and communicated their responses via the ATHENA system. Notable improvements in performance were observed relative to standards achieved at the preceding exercise. The third exercise represented an extremely close approximation to a real-life situation, thus providing an ideal opportunity to learn from the technical performance of the system and also to take account of the various human factors involved in using both the CCCID and the app. For example, it was observed that app using role players at the third exercise frequently did not act as individuals but tended to organize themselves in groups and assumed different roles within the groups. Some of them provided first aid to injured group members, others were seeking for information on the ground to gain local situational awareness and some were using the app to send and receive information.

Overall, throughout the exercises, feedback from both App and CCCID users highlighted various technical and design issues which provided valuable learning opportunities for the future development of the system by conducting them in a dynamic, live-play situation. These issues included such problems as server performance issues and features of the CCCID which only became apparent when the system was exposed to a large number of reports and users.

Analysis of the exercise data also provided an opportunity to accommodate user needs, for example by modifying features such as report templates within the App; due to feedback within from the exercises the need for separate report subject and report descriptions were removed and streamlined into only field. The impact of this may be significant for several other features of the system and so is particularly useful.

12.6   Conclusion

The value of putting the ATHENA system through increasingly demanding testing exercises cannot be overestimated. From a tightly controlled and highly scripted staring point to a large-scale, dynamic live-play exercise where users were spontaneously using the system, the approach adopted was to apply test procedures as robustly as possible while also controlling risks. By such means were we able to record and collect the maximum amount of feedback and results in order to support the system's ongoing development. This process now culminates its most challenging test trial in the context of a 'live-play' exercise, the details of which will now be outlined as we move on to the following chapter.

Chapter 13
The Final ATHENA Test Case: An Integrated View of ATHENA

Alison Lyle, Tony Day, and Kerry McSeveny

13.1 Introduction

The fourth exercise at the West Yorkshire Police (WYP) Carr Gate Complex in Wakefield, UK, represented the final stage of the exercise programme that has been an integral part of the development of the ATHENA system. The aims of the programme as a whole included carrying out a rigorous examination of the system at key stages of its development and maximising its potential, whilst ensuring that the focus on user needs and legal and ethical compliance was maintained. Another key aim from the outset was to derive as much feedback as possible from all stakeholders involved in the exercises and to deploy the system in as many different environments as was practicable.

In furtherance of these aims, the final deployment of the ATHENA system in testing conditions focused on examining whether the system maximised the benefit of the ordinary citizen in a crisis situation, and whether it improved the effectiveness and efficiency of the first responders by facilitating a two-way information exchange and creating situational awareness (see Chaps. 5 and 12).

The final analysis and evaluation of the exercise has not, at the time of writing, been carried out. However, an overview of the preliminary findings emerging from it are offered below, after the exercising method has been described.

A. Lyle (✉) • T. Day • K. McSeveny
CENTRIC, Sheffield Hallam University, Sheffield, UK
e-mail: T.Day@shu.ac.uk; K.McSeveny@shu.ac.uk

© Springer International Publishing AG 2017
B. Akhgar et al. (eds.), *Application of Social Media in Crisis Management*,
Transactions on Computational Science and Computational Intelligence,
DOI 10.1007/978-3-319-52419-1_13

13.2   The Exercise

The Carr Gate exercise differed and progressed from the previous ones in that it was a holistic examination of all the system components working together in a dynamic, real-time environment. The ATHENA App, the Crisis Command and Control Intelligence Dashboard (CCCID) and the Decentralised Intelligence Processing Framework (DIPF, or 'Logic Cloud') were all operated by a range of participants, none of whom were directly involved with the project. Another difference between this and the first two exercises was that the technical testing and validation had been carried out prior to the event. Therefore, the participants were not required to carry out specific tasks to facilitate this. They were encouraged to use the system as if in a real-life situation and as much feedback and data as possible was collected using a variety of methods, to determine the extent to which it satisfied their various needs.

As part of the development of the ATHENA exercise programme, the UK's North East Counter Terrorism Unit (NECTU) created a set of scenarios that were suitable for providing the context in which the ATHENA system could be rigorously tested. The scenarios are fictitious but based on real or foreseeable events and have been created so that flexibility and adaptation is possible. One of these scenarios, entitled 'Public Order Incident' was chosen for this exercise as it had the potential for incorporating all the required elements; it was adapted accordingly by the West Yorkshire Police officers and staff who planned and organised the event.

A separate scenario involving a vulnerable adult and child was also developed; this integrated with the main play but also created an opportunity for part of the exercise to be carried out in various geographic locations, thereby facilitating the testing of the ATHENA Logic Cloud and additional technical capabilities of the system as a whole. This separate 'story' started off in the arena with the main action before moving through the city and out to a rural area several miles away. The interaction and tracking of these participants, and the incoming information from remote groups, added an extra dimension to the proceedings.

13.2.1   Location

The live-play part of the exercise was carried out in West Yorkshire Police's public order training facility, which is situated in a purpose built site. The facility is used to train and tactically prepare officers for a range of different policing situations. Central to this facility is an 'urban area', made up of adjoining indoor and outdoor spaces. One of these spaces takes the form of an indoor arena, comprising a network of marked roadways organised around a central 'public square', flanked by full-sized buildings including shops, residential buildings, a police station, and a local pub. This space can be modified in such a way as to vary the nature of the road layout, and alter the prevailing environmental conditions (e.g. in terms of the amount of light, noise, mist or smoke currently in evidence) in order to make the situation appear as realistic as possible.

Fig. 13.1 Night-time response simulation in the indoor arena

The outdoor arena also includes a road network, full-sized residential buildings, a hospital and multi-storey tower. Both arenas are covered by CCTV, and the indoor arena is overlooked by a viewing gantry.

The role player involved in the separate scenario moved out of the police facility in the early stages of the exercise, and was transported by car to the city centre, four miles away. From there, the role player moved to a different location within the city and finally was taken six miles out of the city centre to a rural location. Additional participants were situated at the various locations and each had a mobile device with the ATHENA App.

Within the public order training building, two CCCID stations were established, each with separate controls and screens and each filtered to receive information from different sources. The first received posts from social media and the second had incoming reports from the ATHENA App. The single, situational picture created on the crisis map was displayed on a large screen on the wall, giving the crisis manager an overall view as the incident developed. The ATHENA Logic Cloud was also established in this location, with separate screens. All the components of the system were set up and tested the day before the exercise, in order to ensure that all elements were working and full connectivity was established and stable.

13.2.2   Participants

Due to the secure environment within which the exercise was carried out, and the policies and practices enforced therein, it was decided to invite all the participants for the exercise from within West Yorkshire Police. These were made up of police support volunteers, police staff and police officers. Although all were members of

Fig. 13.2 CCCID stations within the public order training building

the Force, the range of experience, skills and backgrounds of these individuals afforded the opportunity to gain feedback from a variety of perspectives. Neither these nor any other participant had previous knowledge or experience of the ATHENA system. This was an intentional part of the exercise planning in order to test how intuitive each component was to use; an important consideration for a system designed to be used in extreme and stressful conditions.

Participants were asked to use and test different parts of the ATHENA system. The largest number used the ATHENA App on their personal mobile devices. This group was divided into sub-groups to represent different circumstances that would exist in a real-life situation. A group of around 20 participants represented ordinary citizens, in and around the location of the main incident. A similar-sized group were asked to participate as role players. They were responsible for acting out the developing storyline and creating the life-like scenarios to which the 'citizens' would react.

Prior to the start of the first scenario, all participants at the police site who were testing the app received a briefing which explained the ATHENA project and the aim of the exercise, outlined the role that they would play in each of the scenarios in the exercise, and communicated important information relating to health and safety and WYP's confidentiality policy. A short tour of the training facility was given so that participants could familiarise themselves with the arena space, and ask any questions that might occur with regard to their role and responsibilities.

Participants were also asked to download the ATHENA App, and were provided with the login details required to access the arena's WiFi hotspots. The participants in the 'role player' group were given detailed, laminated instruction cards specifying the character they had been allocated, where they were meant to go, and how they should behave during each of the scenarios. These participants fulfilled a range of roles, including members of extremist groups (described in more detail in Sect. 13.2.3), van drivers, a bus driver, police officers and mounted police. In contrast, those acting as citizens were given specified locations in the arena to position themselves for the start of each scenario, but they were then free to move around the space and were encouraged to respond to the events around them as naturally as possible.

Those engaging with the remote exercises were at two different locations; one involved police officers and staff and the other involved council employees from

Fig. 13.3 Mounted police in the outdoor village during the exercise

both Equality and Cohesion and Emergency Planning departments, thereby providing an insight into relevant user needs. These participants were asked to download the app and given minimal information, so that the way in which they would use it in 'real life' could be observed.

Members of the ATHENA consortium, who attended the event from across Europe and beyond, also participated by using the app to post social media reports. They were also able to follow the proceedings by viewing the crisis map on their mobile devices.

The CCCID required a total of three participants; two operators who received and sent information and a Silver Command, who made tactical decisions based on incoming reports from various sources. Even though ATHENA is designed to support multi-agency response to a situation, this exercise focused on the way in which the police would use the system to support their needs. Therefore, the CCCID operators were a police sergeant and an experienced communications officer and the Silver Command was a senior police officer, experienced in managing serious situations at this level.

The ATHENA Logic Cloud, which is a separate system that integrates with the app and CCCID, was operated by a single participant. In a real-life situation this component may require a greater number of personnel to operate it effectively; the number and variety of specialists required would differ according to the situation. The part played in the exercise by the Logic Cloud was confined to tracking the progress of the vulnerable adult and child using simulated Automatic Number Plate Recognition (ANPR) information, therefore requiring only one operator.

Those operating the CCCID and Logic Cloud were provided with training and briefing on the day prior to the exercise, so that they were familiar with the functionality before the event began and understood the part they were being asked to play.

13.2.3   The Scenarios

13.2.3.1   Background

The background scenario developed by the UK's NECTU was provided to the participants immediately prior to the event to assist them to visualise themselves in the situation and understand the background story. The scenario described government-imposed austerity measures, which had resulted in cuts to vital public services and job losses. As a response to this, trade union and other groups had mobilised a large-scale protest march in the capital, due to be attended by around 750,000 people. The exercise scenarios that were developed from this background took place in a West Yorkshire town on the same day as the planned march, at a time when police resources were likely to be focused on the events unfolding in the capital.

The main focus of the scenarios was the rivalry between two fictional local extremist groups—the 'Extreme Left Party' (ELP) and 'British Rights Movement Wakefield' (BRMW)—who were mobilising around the event to justify and further their aims, participating in social media discussion about the march and posting antagonistic messages to members of the rival faction. These events provided the backdrop for a number of other occurrences on the day, including the disappearance of a vulnerable adult with a baby, and suspicious activity by two men in a van.

13.2.3.2   Scenario One: A Normal Day [Approximately 20 min]

The first scenario began in the indoor arena, in the streets surrounding South Square in the town centre. Participants playing citizen roles were allocated starting points, but then obeyed the instruction to move freely around the area, taking due note of their surroundings as the play progressed. This scenario was deliberately designed to be low-key, to allow time for the participants to familiarise themselves with the space, and gain practice in using the ATHENA App.

During this time, a number of events occurred, all enacted by role-players, in order to test whether the citizens were able to notice what was happening and make any related reports by using the app. Included among these events was the arrival of a group of six ELP members, who began playing a noisy but good-natured game of football on the square, which lasted for the entire duration of the scenario. In addition to this, two other men in an unmarked white van drove around the streets of the arena, while looking out of the window as they passed through the area. The van eventually pulled up in South Square and both occupants got out and briefly looked around before one of them proceeded to retrieve a bag (a holdall with wires visibly protruding) from a wheelie bin outside a nearby building. This man put the holdall in the van, whereupon he and his partner re-entered the vehicle and quickly drove away.

While this activity was in progress, a message was sent to all ATHENA users, asking them for any available information regarding a missing 'vulnerable' adult with a child. This related to the fact that a man pushing a children's buggy had just

Fig. 13.4 ELP members in the Flowerpot Inn

left one of the residential properties on the square before their eyes, made his exit with the pushchair into the outdoor arena, and then departed from the area in a car. This individual's movements formed the basis of scenario five, which will be described in due course.

13.2.3.3   Scenario Two: Conflict [Approximately 30 min]

The second scenario centered on the town's local public house, the so-called Flowerpot Inn. Seated inside the pub, in addition to the landlord and a small number of regular customers, were a group of six ELP members who met there regularly. The latter were behaving in a good-natured but unusually rowdy manner—singing and chanting loudly, as they were awaiting the arrival of a group of members of the Leeds ELP to join them for drinks and a party. It was only a few minutes afterwards that the anticipated guests arrived and began greeting one another enthusiastically. The singing and good-natured shouting continued all the while, and as the commotion further developed a table was accidentally knocked over in the process.

From inside the pub, it was possible to hear the arrival of a group of members of the rival BRMW, at which point the ELP members inside the pub ran upstairs to the first floor of the building in order to obtain a better view of the proceedings. They immediately set eyes on eight BRMW members who were standing outside, armed with bats and sticks, shouting aggressively all the while, and goading the ELP members to come outside to fight them. Eight ELP members responded to this challenge by leaving the pub and engaging in verbal exchange with the BRMW. While no actual physical contact occurred, weapons were nonetheless brandished within what was a notably aggressive and threatening atmosphere. In the midst of this action, one member of the BRMW lifted his clothing to reveal a concealed firearm to members of the opposing faction. This altercation took place in such a way as to block the doorway of the pub, making it impossible for any bystanding citizens to leave safely.

At this point two police officers arrived on the scene, and the disturbance began to calm down. The ELP members then returned inside, and the BRMW ran away, shouting obscenities and threats as they did so.

13.2.3.4 Scenario Three: Bus Attack [Approximately 30 min]

The third scenario centered on a double decker bus located on the east side of the indoor arena, opposite a small police station. The lights in the arena were lowered to create an impression of early evening. The driver and a number of citizens began the scenario by being seated on the bus, while another group of citizens stood awaiting its arrival at a nearby bus stop. The eight BRMW members who had been involved in the disturbance outside the Flower Pot Inn had regrouped by now, and it was they who arrived at the bus stop just as the vehicle was pulling up. The BRMW group mistakenly formed the impression that the citizens on the bus included members of the rival ELP faction arriving from Leeds, and therefore began chanting and shouting aggressively as the bus drew still. They then began issuing threats to the driver and passengers, before attacking the bus with their weapons. In the meantime, the actual ELP members had begun leaving the Flower Pot Inn and had started to make their way across the main square in the opposite direction, towards the northwest side of the town.

The main disturbance continued to progress unchecked. That was until two police officers finally arrived at the scene, started to confront the BRMW members, and attempted to make arrests. In the resulting tussle one officer was assaulted with a weapon and fell, badly injured, to the floor. The second officer tried to call for assistance, but his radio did not appear to be working adequately. The officer therefore turned his attention to trying to support his colleague, who was lying on the ground unconscious and bleeding, but was obviously panicked and distressed due to the fact that he was unable to receive any response from the control room. Soon afterwards, the loud sound of sirens permeated the area, and the BRMW members quickly fled the scene, leaving the two police officers on the ground, surrounded by discarded weapons, one of which was a firearm.

13.2.3.5 Scenario Four: Gas Leak [Approximately 25 min]

The focus of the final scenario of the day shifted away from the conflict between the ELP and BRMW All role-players who had so far been enacting these roles now took on additional 'citizen' roles, and were asked to use the ATHENA App from this point onwards. This newest scenario took place in the outdoor arena, and civilians were positioned at starting points around the outdoor streets and residential areas. Action commenced when an unmarked white van drove slowly around the streets and gradually pulled up outside a house. The van driver and his male passenger then climbed out of the vehicle, shiftily looking around in the process, and set about removing a bag with protruding wires from the back of the van. The two men then placed the bag

Fig. 13.5 Injured officer in scenario three

outside the nearby house and appeared to tamper with the contents before returning to the van. The vehicle was then driven off towards the north west side of the arena, where it was parked in such a way as to be partially blocking the road. Shortly after this manoeuvre had occurred, smoke was seen to be coming out of the bag.

At this point, two mounted police officers arrived on the scene, and begin to evacuate the streets and houses in the vicinity of the bag. The officers told any civilians within hearing distance that the evacuation was due to a gas leak in the area. During the evacuation, however, the CCCID sent out a message to all professional response users that they were under the impression that the bag might contain some device or other that was capable of releasing a deadly nerve agent. In responding to this, the mounted officers then directed those citizens present in the direction of the hospital, which was located on the north side of the outdoor arena. A message was also sent to all citizen users via the ATHENA App, specifying which route they should take in order to bypass the suspicious package. In the meantime, the mounted officers established a cordon in the road, thus preventing people from entering the site of danger and possible toxicity. However, little did they know that the route to the hospital was blocked by the diagonally parked unmarked van, whose driver and passenger had previously been seen leaving the bag outside the nearby house. As the citizens unwittingly approached the van, its two occupants jumped out of the vehicle, while pointing firearms at them, and shouted aggressive threats that they were prepared to shoot anyone who did not move away. This quickly proved to be no idle bluff as a number of shots rang out. It was as this incident unfolded that the vulnerable adult and child (last seen in the arena during scenario one) returned to the outdoor arena.

These events effectively left a group citizens stuck between the police cordon on one side, and the firearm-wielding van occupants on the other. If these citizens returned to the cordon, they would be advised about an alternative safe route, and asked to make their way to the main square in the indoor arena. If the trapped citizens used the ATHENA app to log reports about their situation, then information about the alternative route to the main square would also be disseminated by CCCID via the app.

13.2.3.6   Scenario Five: Vulnerable Adult with Child [Approximately 3 h]

This scenario involved testing the capabilities of the Logic Cloud element of the ATHENA system. After leaving the arena at the start of scenario one, the man playing the vulnerable adult and child, which was a toy doll, travelled by car to a West Yorkshire Police office in central Wakefield. Having entered the building, they then remained for a further 20 min, initially taking the lift to multiple floors, and then walking around office areas, before finally leaving and moving on foot to Wakefield Council's Wakefield One building, where they were to spend a similar amount of time. During their visit to this second location, the vulnerable adult and child first entered the main atrium and proceeded to look around, before eventually arriving at the entrance to the building's café, where they began looking at the menu. The two of them chose, however, not to actually enter the café, preferring instead to leave the building, return to their car and were driven away. The vulnerable adult and child then travelled on to a local National Trust site, where they spent approximately 30 min walking around the public areas and visiting the site's café. Having done all this, the pair then returned to the arena by car, and joined those citizens who were now assembled in the outdoor arena as the events of scenario four were in the process of unfolding. As a central part of scenario five, staff at the various office locations alluded to above were asked to report any possible sightings of the missing person, and regular alerts regarding the licence plate of the car they were travelling in were sent out via a simulated Automatic Number Plate Recognition (ANPR) system.

13.2.3.7   Social Media Activity

For the duration of this entire exercise, ATHENA consortium members, who were present as observers of the event, were cast in the roles of social media users, with a view to generating Twitter content (using the designated hashtag #athenatest) to be processed by the ATHENA system. These observers were provided with a guide to the wider context of the scenarios, alongside a list of specific times at which they were required to tweet particular types of messages related to the relevant events. For example, during scenario one, the social media users were asked to pose as either ELP or BRMW members, sending friendly messages to their fellow group members. Later in the scenario, they were asked at various points to latch onto and reply to a tweet sent by someone from the rival faction, with the objective of kick-starting a potentially antagonistic exchange. Alongside this, the social media users periodically sent unrelated tweets (such as pictures of kittens or baby pandas) tagged #athenatest, to introduce an element of malicious and unrelated reporting on the system and see how these are dealt with. The same users were also asked to play the role of 'concerned citizens', by tweeting to ask for clarification about rumours and seeking advice about what their course of action they should be taking.

13.3   Data Collection

The fourth exercise was attended by a team of eight researchers from CENTRIC, whose aim was to observe the day's events and evaluate the use of the ATHENA system by all participants. Data was collected by use of the following methods: direct observation, capturing proceedings through audio and video recording, a debriefing questionnaire for all role players, and a group interview with the five members of the main police command team. These methods are now described in closer detail.

13.3.1   Direct and Participant Observation

The full exercise (including all preliminary briefings and the actual scenarios) was observed by the entire team of researchers. The individuals concerned were given details of the scenarios a day prior to the exercise, along with a list of suggested vantage points from which to observe events as they unfolded. Two members of the team recorded each scenario with hand-held video cameras, and photographs were taken of key events by a West Yorkshire Police photographer who had been briefed beforehand by the exercise organisers. In addition, each exercise was recorded by the CCTV cameras in the arena, and could be observed by consortium members either on the ground, from a viewing gantry overlooking the indoor arena, or via CCTV from an observation room. An observer was also placed in the 'control room' for the day, to observe the police personnel who were operating the CCCID.

In order to obtain as full and empathetic an insight as possible into the experience of the participants in the exercise, two members of the research team actually participated in the scenarios, one as a 'role-player', playing the part of a member of the ELP, the other as a 'citizen'. These researchers had no prior knowledge of the scenarios, and were therefore able to experience them from a similar perspective to that of the other participants. Each researcher produced a written report of what they actually witnessed on the day, and of any experience they had of using the app while the scenarios were in progress.

13.3.2   Feedback Questionnaire

At the conclusion of the exercise, all those participants playing civilian roles attended a 'hot debrief' session, where they completed a paper-based survey questionnaire which had been designed to elicit their opinions about the ATHENA app, and to generally reflect on their experiences during the exercise. Thirty-six end users were asked 25 questions relating to the following topics:

- Making sense of the incidents during the scenarios (through the use of the ATHENA system)

Fig. 13.6 CCCID report list and crisis map, showing details of banner headline posted by police

- User experience of using the system
- Ethical implications for using the system
- Functional aspects of the system

It should be added that survey responses were supplemented by informal information captured in the course of impromptu exchanges between observers and participants occurring in the course of, or in breaks between, the five scenarios.

13.3.3   Group Interview Session

A corresponding 'hot debriefing' session for senior police officers took the form of a group interview, conducted by the two most senior members of the CENTRIC team, with the five most senior police officers involved in the exercise: the Silver Command, a police sergeant operating one of the CCCID stations, an experienced communications officer operating the other CCCID station and two other police sergeants who had been deployed to deal with the scenario incidents. Here, an opportunity was taken to engage in a discussion about the police experience and evaluation of the ATHENA system, from the perspectives of both field and remote responders.

13.4   Evaluating the ATHENA System

The preliminary findings are significant in relation to several important considerations arising from the use of the ATHENA system. All participants fully engaged with the activities and quickly understood the way in which the system was designed to operate. As a result of this, they were able to provide valuable feedback about their experiences, from a range of perspectives. This process of deploying the system, gathering feedback and feeding lessons learned back into the design and development of all components, has played a crucial part in the iterative, user-centric approach that has been adopted throughout the project. Some of the comments received are presented and discussed below, in order to illustrate this process and the ways in which the ATHENA solution could be used in the future to fulfil user needs in crisis situations.

13.4.1 Technology

From a technological point of view, the CCCID functioned according to plans and expectations. Reports were received from app users via social media as well as text, photo and video reports from the app. These were seen on the crisis map by those managing the situation to understand what was happening and how to respond. The CCCID operators posted reports and instructions to app users and made use of the various features available to them. With a minimum of training, both operators and the Silver Command used the system efficiently and effectively; they were all enthusiastic about the potential of the system and commented on its value as a source of information that other systems don't provide. Due to the speed with which the operators familiarised themselves with the functioning and purpose of the CCCID, they were able to offer valuable feedback from a police point of view.

Both operators commented on the amount of incoming information, from all sources. They acknowledged that during an event with a greater number of people involved, this may be problematic. This is a concern that has been raised previously, and was also anticipated by the project; during a crisis there will be a large number of people sending and requiring real-time information. In recognition of this, an automated report aggregation feature has been developed. This capability will group large numbers of reports together, by subject and location, to present information in a more useful way to operators.

Another concern for the police, as well as the citizens, was the possibility of non-urgent, irrelevant or malicious reports 'cluttering up' the system. The ATHENA system allows for reports to be transmitted more impersonally, making the psychological barrier to sending hoax information easier to cross. The officers identified this as already being a problem for police forces, in respect of conventional communication channels such as phone calls:

.....you've only got to look at recent data, 20 percent of the calls to the West Yorkshire Police customer contact centre are non-police related, and we can't ignore that.'

'And my fear with this system would be, if it's not used like the boss has said, for certain types of incidents, it'd just be overload .... And it'd take a team of people just to work out.'

A feature, that is one of the final ones to be developed, is one that will assist with this issue; automated credibility rating will run using pre-determined values to assign a level of credibility to incoming reports. This will provide an indication to operators on the reliability of the information; this feature has been designed to take into account legal and ethical compliance, relevant to such processes. Additionally, a category labelled 'malicious' is to be used to retain reports that may be of evidential use in any criminal proceedings that may ensue, subject to police data protection compliance.

Some of the problems reported by users of the ATHENA App were of a purely functional nature. Most fundamentally, it was pointed out that, in order for civilians to make most effective usage of the system, they must have access to a phone with sufficient power to support the software and gain access to internet data. Those potential users lacking such technology (there were five such individuals amongst our respondents) were therefore at a disadvantage. The ability of users to download the ATHENA App has been increasingly successful throughout the exercise programme, as the prototype developed. It is of fundamental importance that reliability and stability are fully established in real-life deployment. Out of all the participants very few had problems of this nature. However, this useful feedback, along with the technical data from the day, will provide the app developers with an insight into potential problems:

'The app crashed when I was about to upload a really useful video. Made it feel like I wasted my time doing it.'

'I have a slightly older model of phone and the app didn't seem to be well suited to the small screen –'

Connectivity is also of crucial importance and although every effort had been made during the planning stage of the exercise, some participants lost connection with the WiFi at certain locations, which resulted in the app closing. However, WiFi was chosen for the exercise as an alternative to asking participants to use their own data. In real-life, users would rely on the telecommunications network resilience. In recognition of the importance of connectivity, the project has explored the development of phone to phone Bluetooth and/or infrared transmission that would route a geo-located distress signal to search and rescue services.

13.4.2   User Needs

The almost unlimited positive feedback generated by the survey questions, completed by app users immediately after the exercise, reflected the considerable ease with which people were able to use the ATHENA App. Indicative of this enthusiasm were such comments as: 'easy to log in', 'fast', 'responsive', 'easy to read' and 'easy to log and upload photos and details'. One respondent undoubtedly spoke on behalf of many more in characterising the system quite simply as 'Interesting technology to report incidents as well as get informed on crises as they are unfolding near me'. This latter, sense-making function of the system was regarded as perhaps its most significant and valuable feature and is an indication of the ATHENA vision being realised:

> 'Rather than not knowing there was an "incident happening" near me it was helpful to know what was happening and updates helped me feel reassured…also the opportunity to comment if I had seen something untoward'

> 'Allowed me to recognise the potential risks'

The ATHENA vision also recognises the need for users to feel empowered to help themselves and others. This was expressed by participants who felt that they were making a positive contribution to the common safety and welfare:

> 'Made me feel like I have the opportunity to contribute to my fellow citizens by using the app and being vigilant'

> 'Knowing that we were in contact with the correct people'

It was apparent from the feedback that users felt reassured, which is of great importance in crisis situations, and another of the aims behind the design of the system. Linked to this is the knowledge that their information is contributing to response efforts, enabling those managing the situation to have a well-formulated and relatively accurate impression of the event before arriving on the scene. This gave users a sense of worth and satisfaction:

> 'I felt reassured that everything I reported was being looked at.'

> 'I was confident taking pics/clips as knew it goes directly to the police and would be dealt with.'

> 'It enables me to contribute to police efforts.'

> 'Quick upload of photos which allows authorities to know what to expect when turning up.'

Another important need for all users is for the communications to be clear, unambiguous and useful. On the part of the app users there was a widespread consensus that messages transmitted by the police were suitably intelligible and beneficial:

> 'Clear concise messages well delivered with safety and requests for additional info.'

> 'Information was clear and in language that is easy to understand.'

> 'Helpful if accurate and true. Police updates better than the public's.'

However, the difficulty in achieving this for all users was also evident:

> 'If anything I was confused by the way the information was presented through the app—it wasn't clear exactly what was happening and where it was.'

From a police perspective, the Silver Command on the day referred to the value of having as much information as possible from citizens 'on the ground' so that the correct response could be coordinated:

Fig. 13.7 Different views available on the ATHENA App (see Chap. 7)

'From my perspective as a Commander, I've never used a system like this before where we are essentially providing general broadcasts providing general information to the public, where on this system everybody is able to view that and respond back to it, tasking certain people who have volunteered information appropriately to try and get something back, because constantly all I'm wanting to do, I'm information hungry, in terms of the information intelligence, what do I know, and what do I need to know? And that second box, I'm always trying to fill that, what do I need to know in order to assess threat and risk, and deploy appropriately to that? ....So there's the general broadcast for information..........'

Of particular value to this position was information in the various forms facilitated by the app. For example video footage of the incident allowed a first-hand view to be provided directly to the CCCID operator, and therefore the Silver Command. Respondents took particular satisfaction from the knowledge that they now had the capacity to feed information visually as opposed to just verbally—to the police, thereby enhancing the latter's preliminary grasp of what was actually going on:

'[Ordinarily] I would have just called the police and tried to describe best what's happening—whereas on the app I can record it'

This feature of the app eliminates uncertainty that can arise with the use of written words to report facts, particularly in stressful conditions. A feature of the CCCID that also works on this understanding involves the operator creating a 'danger zone'—an area marked on the crisis map as a visual communication of particular locations for citizens to avoid.

Fig. 13.8 Visual reports gave CCCID operators a powerful insight

13.4.3   Ethical Considerations

Legal and ethical considerations and compliance have been incorporated into the design of the ATHENA system, and included in all the stages of development. An important part of this work has involved adopting the perspective of citizen users in an attempt to anticipate concerns, which may result in a reluctance to engage with the system. It is a central aim of ATHENA that it creates reassurance about both personal and information security. The protection of personal data, as well as upholding ethical values, are topics that the project takes seriously. Despite this, some users voiced their concerns in relation to such issues:

'I don't agree with the app having access to personal information such as social media.'

'I was unaware it was using my personal details.'

Even though the system displays a simplified privacy policy, outlining citizens' rights in this respect, and would only process personal data in a legally compliant way, these comments illustrate the importance of emphasising this and creating a clear message to users.

Ethical considerations relating to providing clear understanding and reassurance to all users, particularly those citizens caught up in a crisis situation, are of utmost importance. As highlighted above, enabling those contributing information to feel that they are helping, as well as being helped, is of both ethical and practical value. Some comments from participants underline this need:

'Not sure if incidents I reported has been received or read. A notification of some kind would be a good feature'

'It's not clear what happens to any of the info I submitted. i.e. I don't know whether that is passed to police etc.'

'Didn't make me feel reassured because I didn't know if anyone could see what I was posting'

Although report receipts and indications that reports have been read are included in the ATHENA system, the need to avoid causing further confusion or worry in an already stressful environment by providing clear indications, is reiterated here.

Another potential use for the information processed by the ATHENA system, is in relation to the detection, investigation and prosecution of criminal activity occurring during a crisis. Police officers were clearly aware of the inherent power of some messages volunteered by the public to possibly incriminate those individuals whose activities were being reported. These messages were therefore seen as a potentially vital resource:

'Don't forget that these are potential witnesses, so members of the public we're seeing as our potential witnesses, and often during the process, we did it two or three times, we sent out a message from the person that had sent something in saying "please, once it's safe to do so, please approach a police officer over what you have witnessed", so those photographs, that video footage is best evidence, isn't it. In the absence of CCTV, which we use extensively, in the absence of police officers' witness evidence, it's valuable evidence you've got there.'

However, this too would be subject to legal constraints, which have been addressed through the work of the project.

Another ethical issue arising from the feedback related to the strong likelihood of certain section of society being indirectly deprived of an opportunity to use the ATHENA system to their advantage. Three users in particular highlighted that certain levels of linguistic competence and literacy skills were necessary to operate the app. For this reason, they and several others maintained that it would be especially difficult for many elderly, disabled or partially sighted people to use the relevant technology. These concerns are echoed in the work of the project, and as such have been included in user consultation sessions and legal discussions, in terms of human rights. The incorporation of reporting methods other than text allow those with limited language ability to effectively communicate. Feedback from users with various requirements, such as partially sighted people, has been sought in order to understand how they too can benefit equally. Five users further pointed out that people without smart phones would not be able to take advantage of the system. This point was echoed by the police in the control room, who expressed concern that the most vulnerable citizens may be among those who don't or can't use this type of technology for communications. These are valid points and ones that have previously been raised during the lifetime of the project. ATHENA aims to harness the prolific use of new technology in supporting effective response and search and rescue actions in crisis situations, which will be of benefit to all citizens.

13.4.4   Additional Considerations

The post-exercise survey that was employed, asked users to respond to questions relating to safety implications arising from use of the app. Some of the responses confirmed that features of the app, that were designed according to user needs, had anticipated these well. The points raised included the user being, or remaining in, a dangerous position while either typing or recording a report. The feature that allows users to edit their location when sending a report is designed specifically so that they can send a report about a different place after reaching safety. Another point also related to the act of sending reports; a user may miss what is going on by looking at their phone. The audio recording is another method of reporting that would avoid this problem.

Other feedback that confirmed the value of ATHENA features came from the Silver Command on the day, who was eager not to cause any unnecessary panic among the citizens, and stressed the importance of ensuring that messages to the public were *'crystal clear'* and the need to *'absolutely nail home the need for them to be safe'*.

The concept of the ATHENA vision incorporates the idea that citizens sending and receiving messages are engaging with the authority that is managing the coordinated response to the crisis in which they are involved. It is important that citizens have trust and confidence in those operating the CCCID. This particular exercise focused on a police perspective, whereas previous deployments have involved other emergency responders controlling the situation. Another part of the post-exercise questionnaire asked respondents whether they knew that the information was posted by the police. The responses indicated uncertainty in this respect. Although in response to a specific question, this indicates that it may be in the interests of creating further reassurance to app users, that the authority managing the crisis is identified. The questions asked in the survey sought to raise particular issues such as this, but also provided an opportunity to collect spontaneous feedback. As has been shown, the comments and insights provided by users are of enormous value and have been so throughout the evolution of the system. In this way, user needs have been, and remain, a central focus of the ATHENA vision.

13.5   Conclusions

It is evident from these brief distillations, both of the enactment of and the feedback relating to, the fourth evaluative exercise at the police Carr Gate training complex in Wakefield, West Yorkshire, that the technology and protocols comprising the ATHENA system now have a proven capability to enhance the confidence and effectiveness of emergency personnel engaged in the difficult art of crisis management.

Whilst it would be disingenuous of us to pretend that the interrelated scenarios described above were likely to approximate in scale and complexity to the 'real life' situations liable to be encountered by the police professionals involved in our exercise, there is no doubt that such officers were sincere when insisting that, even on the basis of a modicum of prior training, they found the ATHENA technology not only comfortable and satisfying to use, but also of great utility in terms of their decision-making activity and two-way communications with the general public.

Our case study actors occupying the roles of the general public were correspondingly impressed by the ease with which they were able to access and utilise the ATHENA Mobile App, and professed themselves delighted by the accentuated sensations of empowerment and reassurance they almost universally experienced. They regarded the opportunity provided by the system to transmit information visually as well as verbally to the authorities as an especially significant and greatly appreciated development.

There were, however, a small number of concerns relating to ethical issues and matters of health and safety. Paramount among these were the felt needs to issue an even more robust set of policy statements regarding the individual's loss of privacy in sending messages to the police, and the fact that the latter could conceivably use (or even follow up) such information in the detection or apprehension of possibly criminal activity. There is now a similar awareness of the need to deter individuals from possibly disregarding their own safety while in the process of creating and transmitting messages. This fine-tuning of the ATHENA system is now in progress in keen anticipation of its application to actual, rather than simulated, crises, emergencies or disasters.

Chapter 14
Concluding Remarks

Babak Akhgar and David Waddington

14.1 The ATHENA Initiative: Raising 'Situational Awareness'

The preceding chapters of this volume have each been geared towards the principal objective of establishing the underlying principles and technological basis of an interactive platform for the management of crises and disasters. We have considered it most expedient and potentially illuminating to develop this discussion in terms of the EC-funded ATHENA project, being carried out according to the technical leadership of the CENTRIC research collective within Sheffield Hallam University and coordination of West Yorkshire Police.

The ATHENA initiative has set its stall out to determine the ways in which it might be possible to utilise web-based social media and mobile devices to improve two-way communication between and within the general public, police and other emergency services, and to enhance the situational awareness of the various parties involved in potential or actual instances of crisis or disaster. Akhgar and Bayerl [1, p. 5] usefully define this crucial concept of *situational awareness* (SA) as 'The capability to identify (people, events, materials, locations, relationships...), contextualise, visualise, process and comprehend the critical elements of intelligence about particular areas of concern. Areas of concern can be anything from an investigation to the management of a major crisis.' By taking ATHENA as our exemplar, we have been able in previous chapters to set out both the communicative requirements and technological composition of an SA platform (whether ATHENA or a similar initiative) for deployment in the context of crisis or disaster. It only remains for us to make a suitably coherent restatement of such imperatives.

B. Akhgar (✉) • D. Waddington
CENTRIC, Cultural, Communication and Computing Research Institute (C3RI),
Sheffield Hallam University, Sheffield, UK
e-mail: B.Akhgar@shu.ac.uk; D.P.Waddington@shu.ac.uk

© Springer International Publishing AG 2017
B. Akhgar et al. (eds.), *Application of Social Media in Crisis Management*,
Transactions on Computational Science and Computational Intelligence,
DOI 10.1007/978-3-319-52419-1_14

14.2 Communicative Requirements

Our introduction to this volume not only highlighted the accelerating growth and pervasiveness of the use of social media in all modern societies, but further argued that it was essential to fully utilise this 'Information Age reality' in such a way as to animate, involve and empower the members of these societies by engaging them in the sharing of information (and SA-raising activities) pertaining to crisis and disaster. It was emphasised in the first of two contributions by Kerry McSeveny and David Waddington (Chap. 2) that the police and other authorities have traditionally been reluctant to engage the public in this manner, due to erroneous misconceptions that lay individuals caught up in crises are prone to act in fundamentally panic-stricken, irrational and selfish ways. These authors provide a much-needed corrective to this line of thinking by pointing to a strong body of evidence which shows that people are more apt to behave in an orderly, rational and prosocial fashion, and by echoing the general academic argument that the public should be seen as part of the solution, and not part of the problem.

Another important function, not only of this opening chapter by McSeveny and Waddington, but also of their subsequent chapter and a related contribution by Eric Stern (Chaps. 3 and 4), is to provide a number of important prescriptions for ensuring effective two-way communication (and sense-making) between the public and authorities in times of crisis and disaster. Put briefly, these authors exhort the relevant authorities to:

- Try to empathise with the prevailing physical and psychological needs of the public at the time in question
- Establish two-way processes of communication and be intent on listening to and understanding what the public might have to say
- Give due consideration to the possibility that some people may not have access to certain forms of communication and be prepared to compensate accordingly
- Be suitably aware of possible cultural differences between relevant sections of the public (e.g. in terms of relevant values or use of language)
- Inspire confidence and trust by using credible sources to transmit messages (e.g. reputable and familiar individuals or local radio announcers)
- Adopt a suitably concerned, compassionate and empathetic tone (in contrast to the brusque 'command and control' style that is sometimes associated with the police and emergency services)
- Transmit clear (i.e. explicit and unambiguous) information in such a way as to advocate (and explain the underlying reasons for) particular courses of action
- Provide a global overview of events so that people are fully 'in the picture'
- Acknowledge the full seriousness of the situation while striving not to create any unnecessary alarm
- Try to appeal to a collective (or communal) spirit rather than encouraging individualistic forms of behaviour

The above prescriptions relate, of course, to only one direction in the envisaged two-way process of communication exchange. Of equal relevance to our discussion is the anticipated flow of information provided by the public to the relevant authorities. Simon Andrews makes the especially important point (in Chap. 5) that, whilst this type of information is of undoubted practical value, it nonetheless carries the potential to be dangerously counterproductive. Such danger stems from the fact that the sheer volume of communication being received may well result in 'information overload', making it notoriously difficult to prioritise messages in terms of their urgency or importance. The extent to which information may be regarded as entirely credible (i.e. accurate and truthful), rather than bogus or deliberately misleading, is also a matter of great significance to those in positions of authority. Not surprisingly, it has been the aim of those working on the ATHENA project to offset or mitigate as far as possible all potential drawbacks of this nature.

14.3 Technological Components

Every possible attempt has been made to incorporate the above communicative requirements into the ATHENA technical system. As part of his discussion in Chap. 5, Andrews makes the point that any commitment to SA platforms like ATHENA should not be regarded as an attempt to supplant or supersede more conventional channels of communication. He further remarks, however, that there has been a growing uptake of 'social media-based crisis management platforms' by a wide variety of LEAs and NGOs in such contrasting locations as North America, Western Europe and the Middle East.

It is in keeping with these diverse initiatives that ATHENA has been geared towards the development of automated techniques capable of locating, filtering and packaging information of pertinence to a crisis situation, to help crisis managers to analyse, understand and act upon a potential snowstorm of relevant detail, and to allow beneficial intercommunication between and among the authorities and general public.

There is little to be gained by simply restating the various ways in which these technological imperatives have been achieved as part of the ATHENA project. It will suffice to say that the raft of technological innovations described in Chaps. 5–9 inclusive has included myriad filtering mechanisms which automatically help to reduce information overload while making 'priority' and 'credibility' assessments of incoming data. The complexity of the situation is also greatly simplified by having 'map-based interfaces' capable of providing 'geo-visual views' of the unfolding crisis. Meanwhile, the provision of the ATHENA mobile application (or app) enables members of the public not only to locate precisely where and to what extent the crisis is occurring but also to appeal for help and/or post verbal and visual reports for the benefit of lay and official audiences.

The ways in which such mechanisms have been designed are a reflection of stated user requirements, but the underlying principles and techniques on which they are based are universally available, and therefore capable of being re-customised in line with the specific needs and objectives of other potential users.

14.4 Legal and Practical Considerations

Two related contributions to this volume, Chap. 10 by Alison Lyle and Chap. 11 (by Alison Lyle and Fraser Sampson), relate to the inevitable legal tension between the needs to promote citizen safety and security on the one hand, while protecting their personal data and entitlement to privacy, and further examples of human rights on the other. Lyle points out, for example, that methods of data collection and processing inherent to the ATHENA system are not only vulnerable to charges of 'profiling' and 'discrimination', but also in direct contravention of the privacy policies subscribed to by personal providers. Her corresponding chapter with Fraser Sampson dwells on the alarming possibility that the usage by LEAs of ATHENA Big Data may be regarded as a 'covert and unregulated' extension of State surveillance.

The two final substantive chapters of this edition (Chaps. 12 and 13) reflect on the interim and 'final' test case scenarios undertaken in assessment of the feasibility and effectiveness of the ATHENA platform. The first of these contributions reports on the ways in which technical deficiencies were highlighted and duly tightened up or eradicated. What is perhaps most notable of all about the feedback received in the wake of the 'Carr Gate' test case is that, whilst 'members' of the police and public alike were generally impressed by the technological and communicative prowess of the ATHENA system, there was evidence of some concern among the latter with regard to certain legal or ethical issues.

Respondents were evidently discomfited to learn that the app provided access to personal data, with some arguing that the ATHENA privacy policy needed to be far more explicit in saying that, by sending messages to the police, the individual was effectively surrendering their right to privacy. Similar feelings of alarm surrounded the fact that the police might use information volunteered by the public as 'evidence' in the process of crime detection.

The trick, as Lyle and Sampson rightly maintain, is to strike the right balance between respecting the personal privacy of the individual while doing as much as possible to guarantee the safety and security of the wider population. These authors quite reasonably insist that ATHENA and other SA platforms of this nature have a clear need to operate according to transparent sets of guidelines and protocols for the use of personal data, which should ideally incorporate an End User Agreement License. It is, they believe, only by resorting to such safeguarding mechanisms that the likes of ATHENA will be able to maintain the legitimacy and integrity it requires to guarantee the ideal levels of public consent and cooperation.

Reference

1. Akhgar, B., & Bayerl, P. S. (2015). *Situational awareness and intelligence platform*. Briefing to SAP Executives in Walldorf, Germany, 7 September 2015.

Index

© Springer International Publishing AG 2017
B. Akhgar et al. (eds.), *Application of Social Media in Crisis Management*,
Transactions on Computational Science and Computational Intelligence,
DOI 10.1007/978-3-319-52419-1

Printed in the United States
By Bookmasters